SCIENCE OF BALL LIGHTNING (FIRE BALL)

SCIENCE OF BALL LIGHTNING (FIRE BALL)

Tokyo, Japan 4-6 July 1988

Editor

Yoshi-Hiko Ohtsuki

Japanese Information Center of Ball Lightning (Fire Ball)

World Scientific

Singapore • New Jersey • London • Hong Kong

Published by

World Scientific Publishing Co. Pte. Ltd.,
P O Box 128, Farrer Road, Singapore 9128
USA office: 687 Hartwell Street, Teaneck, NJ 07666
UK office: 73 Lynton Mead, Totteridge, London N20 8DH

Library of Congress Cataloging-in-Publication Data

International Symposium on Ball Lightning (Fire Ball)
(1st: 1988: Waseda University)
Science of ball lightning (fire ball).

1. Ball lightning—Congresses. I. Ohtsuki, Yoshi-Hiko, 1936- .
II. Title.
QC966.7.B3I57 1988 551.5'634 89-9004
ISBN 9971-50-723-4

Printed in Singapore by JBW Printers & Binders Pte. Ltd.

Organizing Committee

Prof. Dr. Y. H. Ohtsuki (Chairman)
Prof. Dr. N. Kitagawa
Prof. Dr. H. Kikuchi

Advisors

Dr. S. Singer (USA)
Prof. Dr. B. M. Smirnov (USSR)
(Other 8 Scientists)

List of Sponsors

Anritsu Corp.
Futami Shobo Co Ltd.
Fuji Electric Co Ltd.
Gakujutsu Tosho Shuppan Co Ltd.
Honda Motor Co Ltd.
Jikyo Shuppan Co Ltd.
Hitachi Ltd.
Kyoritsu Shuppan Co Ltd.
Maruzen Co Ltd.
Murata Mfg Co Ltd.
NEC Corp.
Nippon Gaishi K K
NOK Corp.
Otowa Electric Industry Co Ltd.
Sankosha Co Ltd.
Sony Corp.
Toagosei Chemical Industry Co Ltd.
Toa Nenryo Kogyo K K
Yokogawa Medical System Co Ltd.

PREFACE

This volume consists of papers presented during the First International Symposium on Ball Lightning (Fire Ball) held at Waseda University, Tokyo on July 4–6, 1988. It was attended by 54 scientists from 8 different countries. The scientific program of the Conference consisted of 17 lectures.

Ball Lightning is most probably the rare natural phenomenon on earth. Its physical nature is still unknown despite the attempts that have been to understand its guiding principle. Up till now, the experimental simulation of Ball Lightning has been unsuccessful. Many scientists in different fields are interested in Ball Lightning and performed research works in different kind of manner. However, there have not been any systematic organization or conference for these scientists in the world.

Our symposium was the First International Symposium on Ball Lightning, where many informations of eyewitness, photographs and video reports were presented and discussed for the analysis of the informations. Many experimental and theoretical model for Ball Lightning were reported.

The committee and participants are indebted to a number of people outside the field who made the Symposium possible. Especially advance planning was skillfully guided by Mr. Ara and Mr. Kagawa. Finally, we gratefully acknowledge the financial support of the Japanese Industrial Sponsors listed on the following page.

In the process of the editorial works of this Proceedings Prof. N. Kitagawa have helped us so much, whose contribution is deeply appreciated.

Y. H. Ohtsuki

CONTENTS

SCIENCE OF BALL LIGHTNING (FIRE BALL)

Ball Lightning - The Continuing Challenge

J.D. Barry

Hughes Aircraft Company

Mail Code: EO/E51/B221

P.O. Box 902

El Segundo, California 90245

U.S.A.

and

Stanley Singer

Athenex Research Associates

381 South Meridith Avenue

Pasadena, California 91106

U.S.A.

One hundred fifty years ago the first scientific paper on ball lightning was published by F. Arago heralding a new era of scientific inquiry and understanding of this natural event (1). The relationship of ball lightning to natural electricity was suggested by Arago in a larger discourse almost twenty years later (2). It was to be over seventy-five years before the large numbers of reports and observations were widely recognized and ball lightning became the subject of a monograph (3). A second monograph followed in a decade (4). More information accumulated, interest increased, and modern scientific methods, both theoretical and experimental, were directed to this natural phenomenon. A third monograph summarizing the scientific inquiry appeared some fifty years after the first and was followed a decade later by three others (5-8). Ball lightning has thus been the subject of six monographs, over five hundred scientific papers, and the cause of over two thousand reports, many of recent occurrence.

In spite of the extent of this examination, a widely accepted explanation of ball lightning has remained elusive. In addition, skepticism concerning the existence of ball lightning has been expressed repeatedly. Soon after Arago's work appeared, M. Faraday suggested that because of its long lifetime ball lightning cannot be a discharge of atmospheric electricity (9). Others have proposed that ball lightning is an optical illusion, after image, or incorrect identification of other atmospheric objects (10, 11). Such skepticism is to be expected as there is limited unambiguous physical evidence for the phenomenon. Investigations have necessarily been largely concerned with the study of observational data containing large variances and the evaluation of experiments with unproven relationship to the natural event. In view of the skepticism which has diminished but not disappeared in recent consideration, it is necessary to assess the significance of the phenomenon. Certainly, the repeated observation in nature of a striking occurrence which has not been clearly explained demands a serious interest. The attention to this subject as indicated above has been almost comparable to that given any major question in physics and meteorology.

The characteristics of the natural atmospheric fireball have been deduced from the numerous published reports obtained over several centuries with most of the observations made in Europe. The globe is easily recognized as a distinctive object with little resemblance to other atmospheric events. However, because of its random and possibly rare appearance compared to other objects seen in the lower atmosphere (such as ordinary lightning, for example), no measurements have been made with the instruments which have provided significant information about the other phenomena. The data are limited to properties which can be observed by usually unprepared observers, such as size, velocity, color, and lifetime. The qualitative estimates of these properties have provided approximate averages which fit the majority of observations although, in fact, a wide range of properties has been reported.

The available data from Europe suggest an association with very strong thunderstorms and especially numerous or intense flashes of ordinary lightning. The rate of appearance of the ball is far less than linear flashes, however, perhaps 3×10^{-11} km^{-2} sec^{-1} compared to 3×10^{-8} $km^{-2} sec^{-1}$ (12). A few reports describe formation of the ball at the side or at the very lower tip of a luminous, descending lightning discharge. The ball is usually orange or somewhat red in color with a diameter of 30 cm. It may move through the air with a moderate velocity of about 2 m/s, occasionally entering dwellings. A few of the globes have come close to observers, even touching them. In some of these cases an unexpected absence of heat was reported. In others the observer was burnt. Combustible materials have been ignited, substances melted, and structures damaged by the violent explosion sometimes reported as the ball disappeared (5, 6).

Considering just these most common properties, a wide variation has been noted in individual eye-witness reports. The occurrence of ball lightning in clear or fair weather is unusual in most European reports. In the recent collection of observations from Waseda University, however, cases not associated with storms outnumber those in rain or storms by more than ten-fold. This difference is significant to theories of the formation of the ball, in fact crucial for some which involve lightning processes directly, and illustrates the importance of the observations in this problem. In addition, colors other than orange or red-orange have been noted, including blue, green, and yellow as well as intense, blinding white. Some cases involve a change in color during the observation, and a few balls exhibited a violet-blue corona.

The reported properties have a wide range of values. Reported diameters range from about one centimeter to several meters. Figure 1 shows the values from several collections of observations. We have the most data for this property of ball lightning. It appears that a wide range of values has been reported, even taking into account the difficulty of estimation by eye-witnesses at different distances from the object. In general, the smaller dimensions have been reported for balls closer at hand but not without exception.

The shape is sometimes an egg-shape or a ring rather than a ball. Crackling sounds and odors may be found, possibly similar to those noted with electric discharges. The ball is occasionally stationary, or it may roll or bounce along the ground, or it flies extremely rapidly through the air. Although it is most commonly observed near the ground, ball lightning has been seen formed at some

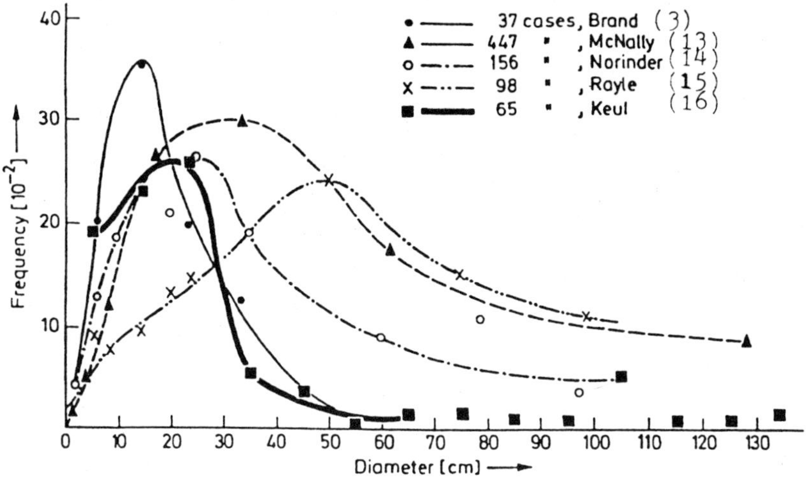

Figure 1. Ball Lightning Diameters.

height in the sky from which it descends rapidly. Sometimes it has been seen rising quickly from near the ground after an ordinary flash of lightning.

Estimates of the energy content of ball lightning have been made from the few incidents with sufficient effects reported to allow such calculations. The estimates range from 5×10^{-5} Jcm^{-3} (17) to over 4×10^{5} Jcm^{-3} (18). An energy density of 5×10^{-3} Jcm^{-3} from metastable molecular excitation or chemical combustion reactions would be sufficient for perceptible luminosity (19). The values of the energy density of ball lightning deduced from over twenty reports are shown in Figure 2 (20).

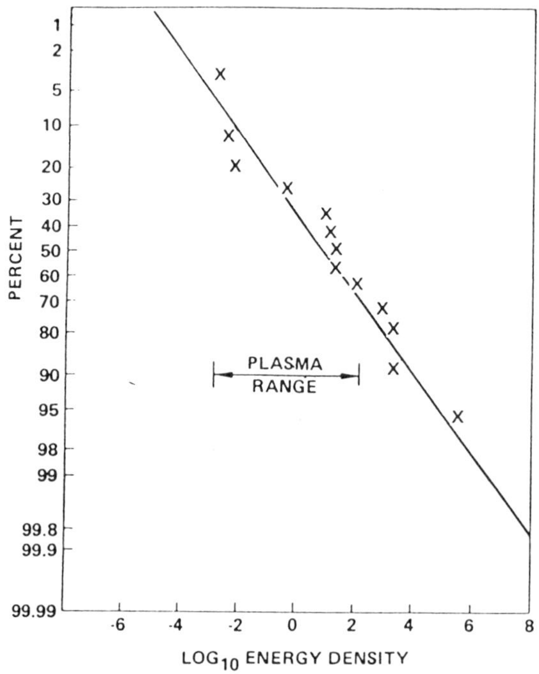

Figure 2. Energy Density of Natural Ball Lightning (20).

The wide range of characteristics found in ball lightning reports inspired the comment that although there are many natural phenomena which are hard to understand, not many become more difficult to explain with more observation (21). Nevertheless, despite the great variation among individual reports, the observed properties indicate a distinct phenomenon.

While statistical measures of ball lightning properties have been used to summarize the results of surveys containing large numbers of reports, the more distinctive but widely accepted characteristics of ball lightning have been derived from the repetition of eye-witness accounts. The review of a few individual reports will best give the flavor of this startling occurrence. Four incidents have been selected here as representative of the many available in the literature. A bibliography of about two thousand entries is available (6).

Event 1. During a thunderstorm a small bright ball the size of a golf ball floated through a door into the kitchen of a house where the observer was standing. It progressed at a leisurely speed, bobbing slightly as it moved to the back of the room where it passed over a stove. The ball then turned, making a horseshoe curve back to the center of the room, still at a man's height. Then it floated out through the door without leaving any permanent trace. The whole incident took place in about one-half minute (22).

Event 2. A white-yellow globe slightly larger than a tennis ball appeared in a restaurant in the Austrian Alps (23). The incident occurred during a thunderstorm with heavy rainfall. The ball was just above the floor as if it had come in through the open doorway at the entrance. The ball floated slowly about two meters toward a St. Bernard dog lying in the entrance hall. The dog jumped away when it saw the ball, and there was a loud bang like a gunshot. Five people who had been in the entrance saw and heard the incident. After the explosion patrons of the restaurant came out through a door from an adjacent room which had been closed and inquired about the noise.

Event 3. During a violent thunderstorm in England several houses were struck by lightning. In one which was not, an eyewitness in the kitchen saw a 10-cm diameter ball appear over the stove (24). The ball was blue-purple with a halo the color of a flame. The ball made a sound like a rattle and gave off intense heat. It moved straight toward the witness and struck her dress approximately 95 cm from the floor, burning a 4 X 11 cm hole in the fabric. She brushed it away with her hand, and it exploded. Her hand became reddened, as did her legs. The whole incident took place in a very short time, perhaps a second.

Event 4. Ball lightning occurred inside the cabin of a commercial passenger airliner soon after it encountered ordinary lightning (25, 26). The passenger who observed the ball, a scientist, reported that seconds after an external flash surrounded the airplane a luminous globe appeared coming from the pilot's compartment down the aisle of the passenger cabin. The sphere was approximately 22 cm in diameter and blue-white in color. The visible luminosity was evidently low, as the light was estimated at 5-10 W. The circumference was slightly darker than the central part of the solid appearing object. The ball appeared spherical, so much that it was not possible to tell whether it was rotating. No radiation of heat was noted. The globe floated straight down the center of the aisle approximately 75 cm above the floor at the commonly reported slow velocity of about 1.5 ms^{-1}, passing the observer at a distance of about 50 cm.

In addition to the information derived from the reports of eyewitnesses, many photographs have been published as records of the appearance of ball lightning. The most complete compilation of these contains over fifty photographs (6). Most of these were obtained by the method common in lightning photography over many decades: a camera was directed at the sky, and its shutter was left open during the storm. The film was changed periodically depending on the frequency of lightning during the storm. The change might occur hourly or even less often, but the camera was unattended for most of the time. Later, the film or plate was examined for unusual images, and those which showed either round objects or, more often, complex traces were classified as ball lightning photographs. As photographic studies of ordinary lightning progressed, however, it became clear that the normal lightning discharge in the atmosphere may display very complex paths. This is in accord with the processes known to occur, from dart leader to the final bright positive current pulse in the ordinary lightning discharge. Thus, the designation of a photograph as ball lightning on the basis of the appearance of the picture alone is generally unjustified. However, a videotape has been obtained showing a fireball formed alongside the discharge channel of ordinary lightning (27).

If a supporting eyewitness account is a required test of validity of a ball lightning photograph, three pictures of all those published remain for serious consideration. The photograph of Figure 3 exhibits a light path reported by two observers to be that of a ball lightning in spite of the camera motion which lead to the blurred images (28). Figure 4 shows a fireball emitting rays, a process noted in other reports of ball lightning (29). The ball was observed during a thunderstorm and was reported to have an upward path of flight. It appeared to leave a residue as it passed. A drawing of a similar lightning ball was published eighty years earlier (30). Figure 5 shows a single image of a fireball which also displayed motion past a building near the observers (31).

Figure 3. Path of Ball Lightning Flight (28).

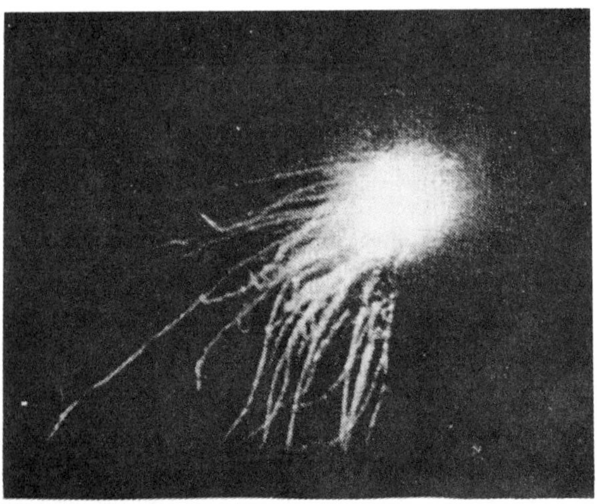

Figure 4. Ray-Emitting Ball Lightning (29).

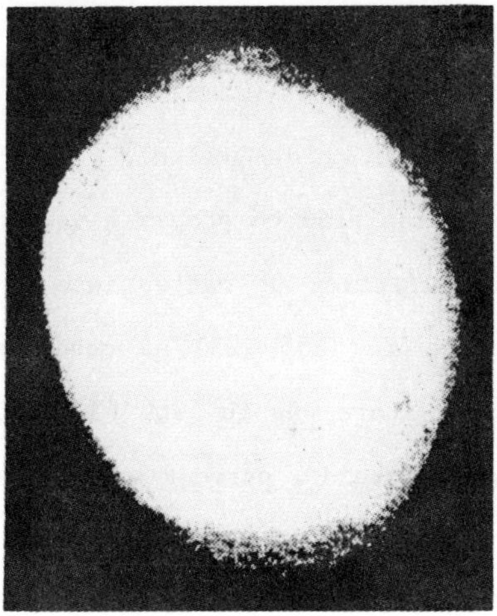

Figure 5. Oval Ball Lightning (31).

The photographs published as being snapshots of ball lightning have been questioned and subjected to skepticism denying their validity (6,32). Past experience in this aspect of the field shows that the support of eyewitnesses of evident probity is necessary to verify ball lightning photographs, but even direct support of this type has not been accepted in all cases.

Skepticism concerning the existence of ball lightning is still expressed regarding photographs and eye witness reports alike. The absence of a conclusive theory for the phenomenon might be cited to account for this situation as well as the conflicting nature of the observations. But there is no lack of theories for this atmospheric object in the scientific literature. In the decade from 1971-1980 alone an average of five papers on ball lightning, usually with new proposals, were published per year (22,32).

The theories range, in general, from relatively modest applications of basic principles to proposal of entirely new states of matter. The natural collection of matter into a spheroidal state no doubt was as reasonable in 1855 (33) as condensation into a dense, metallic superconducting state was in 1966 (34). On the other hand, a theory recognizable as a toroidal current was proposed as early as 1905 (35) and nuclear reaction from high energy particles accelerated by atmospheric potentials in 1928 (36). In spite of the wide range of explanations proposed for the phenomenon, none of the ideas has been able, as yet, to account for the accepted properties of the object.

The properties which appear to present the greatest difficulty among those reported are: (a) the extended period over which the ball remains luminous in many reports, (b) both the regular motion in some fireballs and the irregular motion in others over an extended path, and (c) the sudden appearance of a number of events in clear weather or, at least, in the absence of marked atmospheric electrical activity. A few theories account very well for specific properties which were selected for major consideration. One would hope that some minimum level of comparability with the reported properties of ball lightning be observed in future theoretical work.

Experimental attempts to reproduce ball lightning in the laboratory have also had only limited success. Three types of glowing spheres have been produced which bear close resemblance (to the extent that they have been investigated) to the observations of ball lightning in nature. Studies of the experimental phenomena have not progressed far enough with any of the three to give a complete picture of their properties. In addition, as with the theoretical models, none of the experimental fireballs exhibits all of the reported properties of the natural ball under proper atmospheric conditions. Some differ in one or two specific characteristics (32,37).

The greatest recent attention has been given to plasma assemblies, or plasmoids, which are commonly under study in connection with plasma confinement for thermonuclear fusion. Plasmoids may be considered related to the electrical discharges long of interest in ball lightning studies. Both ac and dc atmospheric discharges have produced luminous globes with some resemblance to ball lightning including reasonable lifetime, but thus far only in laboratory vessels (5,6,38-42). Interest in ball lightning arises almost universally among scientists in plasma studies dealing with geometries such as the spheromak. An essential result has been found for axisymmetric toroidal discharges indicating that after an initial unstable period the plasma relaxes into a quiescent, stable state (43).

Another fireball may be produced by ignition of dilute fuel gases in air such as hydrogen, methane, propane, and benzene (19). Low concentrations of one to three per cent, generally below the usual combustion limit, give the effect. The spherical glow travels for some distance but only in the laboratory vessel containing the uniform dilute mixture. This fireball is another example of an apparently simple phenomenon which has not been explained. The published work has dealt with the formation and qualitative visible characteristics of the globe but has not extended to basic consideration of the stable ball size or the processes responsible for the light emission. According to present combustion theory, ignition of a fuel-air mixture within the combustion limits would normally give a flame front progressing spherically from the point of ignition.

The appearance of a liquid fireball has been reported in combustion of a solid mixture of tetrazole and sodium tetrazolate(44). A light red ball 1-2 mm in diameter forms at first, growing to 13-17 mm as the combustion progresses. However, an estimate of 0.1 gcm^{-1} was obtained for the density of the ball, which is much higher than that of the gases but lower than that of the liquid. The investigators attributed the intermediate density to gas bubbles in the fireball.

The third type of luminous ball is the one produced by a strong discharge of electricity through iron, copper, and lead wires (45). The glowing, mobile spheres float in the air and bounce off surfaces they happen to encounter. A few instances are known from accidental observations recorded in the literature, but these globes have been investigated only to a limited extent.

It is evident that the three types of fireballs observed experimentally have been studied insufficiently to determine whether they are related to the globes reported as ball lightning. Further examination of these laboratory balls may clarify many aspects of the natural spheres.

Our understanding of the ball lightning phenomenon has progressed significantly since the first serious investigations in the early 1800's, yet we lack knowledge of the mechanism of formation and the structure. It appears that further progress in understanding ball lightning will be made only by adequate consideration of rational physical principles and experimental study of reported phenomena which resemble the natural object in certain obvious characteristics.

REFERENCES

1. F. Arago, Annuaire, Bureau des Longitudes, 1838.
2. F. Arago, Oeuvres Completes, Claye, Paris, 1854. Vol. 4.
3. W. Brand, Der Kugelblitz, Henri Grand, Hamburg, 1923.
4. M.E. Mathias, La Foudre et Sa Forme Globulaire, L'Office Nationl Meteorologique, Paris, 1935.
5. S. Singer, The Nature of Ball Lightning, Plenum Press, New York, 1971.
6. J.D. Barry, Ball and Bead Lightning, Plenum Press, New York, 1980.
7. I.P. Stakhanov, The Physical Nature of Ball Lightning, Energoatomizdat, Moscow, 1979.
8. B.M. Smirnov, The Question of Ball Lightning, Nauka, Moscow, 1987.
9. M. Faraday, Experimental Researches in Electricity, Vol 1, Bernard Quaritch, London, 1839. p. 523.
10. E. Argyle, Nature 230, 179 (1971).
11. W.M. Charman, J. Atmos. Terr. Phys. 33, 1973 (1971).
12. B.N. Turman, J. Geophys Res. 82, 2566 (1977).
13. J.R. McNally, Jr., Bull. Am. Phys. Soc. 6, 202, Abstr. J-14 (1961).
14. H. Norinder, Kgl. Vet. Arsbok 94, 39 (1939).
15. W.D. Rayle, NASA TN-D-3188, National Aeronautics and Space Administration, Washington, D.C., January 1966.
16. A.G. Keul, Naturwiss. 68, 134 (1981).
17. P.W. Burbridge and D.J. Robinson, Nature 300, 623 (1982).

18. M.T. Dmitrev, B.I. Bakhtin, and V.I. Martynov, Sov. Phys.-Tech Phys. 26, 1518 (1981).

19. J.D. Barry, J. Atmos. Terr. Phys. 30, 313 (1968).

20. J.D. Barry, J. Geophys. Res. 85, 4111 (1980).

21. F. Sauter, Meteorol. Z. 12, 241 (1895).

22. S. Singer, "Ball Lightning," in "Lightning," R.H. Golde, ed., Academic Press, New York, 1977. Vol. 1, pp 409-436.

23. G.A. Wagner, Nature 232, 187 (1971).

24. M. Stenhoff, Nature 260, 596 (1976).

25. R.C. Jennison, Nature 224, 895 (1969).

26. R.C. Jennison, Nature 245, 95 (1973).

27. A.J. Eriksson, Nature 268, 35 (1977).

28. H. Petersen, Geophysica 5, 49 (1952).

29. E. Kuhn, Naturwiss. 38, 518 (1951).

30. W. v. Haidinger, Sitzber. Kgl. Akad. Wiss., Wien, Math.-Naturwiss. Kl. II 58, 761 (1868).

31. K. Schneidermann, Feuerwehr-Verbands-Z. 44, 1 (1934).

32. S. Singer, Naturwiss. 67, 332 (1980).

33. A. Poey, Compt. rend. 40, 1183 (1855).

34. G.S. Teletov, Priroda 55, No. 9, 84 (1966).

35. V. Carlheim-Gyllenskold, Ber. Int. Met. Direktorenkonf., Innsbruck, 1905; K. Z. Met. Geodyn. Anh. Jahrbuch, 1905.

36. K.F. Bottlinger, Naturwiss. 16, 220 (1928).

37. C. deJans, Ciel et Terre 31, 499 (1910); 32, 155, 255, 301 (1911); 33, 18, 143 (1912).

38. G. Plante, Compt. rend. 80, 1133 (1875).

39. M. Toepler, Ann. Phys. IV 2, 560 (1900).

40. S.A. Arrhenius, Lehrbuch der Kosmischen Physik, Hirzel, Leipzig, 1903. Vol. I, p. 772.

41. P.L. Kapitsa, Dokl. Akad. Nauk SSSR 101, 245 (1955).

42. J.R. Powell and D. Finkelstein, Am. Scientist 58, 262 (1970).

43. J.B. Taylor, "Relaxation of Toroidal Discharges," in "Proc. Third Topical Conf. at Culham Laboratory," Pergamon Press, Oxford, 1976. pp 59-67.

44. A.I. Lesnikovich et al, Nature 323, 706 (1986).

45. D. Van Marum, Phil. Mag. 8, 313 (1800).

HUNGARIAN BALL LIGHTNING OBSERVATIONS IN 1987

George Egely
Central Research Institute for Physics
H-1525 Budapest 114, P. O. Box 49, Hungary

Abstract

Nine ball lightning observations, which were recorded in 1987, are described briefly. In three of these cases the estimated total energy content of the phenomenon exceeded 10000 Joules. In an another case, long distance electrical effects were observed in a meteorological observatory. In four additional cases the spontaneous formation of the phenomenon is described. A brief statistics is provided on the temporal and spatial distribution of the observations.

Ball lightning (BL) is a rare natural phenomenon and its fundamental principles are still not understood. As the phenomenon cannot be simulated in laboratories, only the collection of detailed observations could lead to the deeper understanding of it.

The systematic collection of the observations started in 1986 and after an initial screening over 400 of them were included in the written collection, over 250 of which have already been published[1)]. 24 observations were collected in 1986 (when there was rather dry summer), and 36 were collected in 1987. In most cases the ball lightning caused no or only minor damage; however, there were some cases where the destruction was rather severe.

The phenomenon has been observed during thunderstorms in 18 cases, in cloudy but not stormy weather in 13 cases and in good weather without clouds in 5 cases. In the rest of the cases the weather was not recorded.

The monthly distribution of the observations was the following:

April 2; May 4; June 14, July 13; August 2; September 2; October 1; November 1.

Most of the ball lightning were observed when cold fronts moved through the country causing thunderstorms.

There were 3 observations between 16th and 17th of June, 4 observations on 27th of June, 6 observations between 24th and 27th of July and 2 observations between 2nd and 3rd of August.

1. Mrs. Némethe reported the following observation on their goose farm at Fábiánsebestyén on the 27th of June (case 321).

"A storm passed and the weather was already calm, cloudy at about 8 p.m. Suddenly I spotted a floating, grayish spherical object floating above the brook. It had a diameter of about 2 meters. It was about 30-40 meters from us and moved towards our building with the speed of a running man.

I told my husband to come and look at it, but in 2-3 seconds, by the time he reached the window, we heard a loud crash above our heads, and the 35x15 meter roof of the building was lifted off and fell onto the neighbour's oat field.
The brick chimney and its 1.5 meter long iron extension fell and the back wall of the building (single 12 cm thick bricks) collapsed. There was no trace of fire or heat. A girl at the neighbouring farm (about 100 meters away) saw a reddish ball moving away from the ruined house right after the loud crash. It floated just above the oat field and moved horizontally."

The owners noted a strange thing: a steel reinforced concrete pole broke at ground level, pointing towards the building after the events. The average weight of the roof (corrugated slate, heat insulation, wooden pylons) is estimated to be about 20 kg/m^2, and it was lifted at least 1 meter according to the withnesses. Therefore the total work was abut 10000 Joules, but the energy density cannot be estimated because the BL floated away.

2. The second destructive case was reported from Sátoral-

jaujhely by Mr. and Mrs. Vince (case 391).

"On the 25th of July the weather was cloudy but calm. Several people were working at different construction sites in the morning. At one house a man suddenly noticed a small yellow luminous ball, which slightly singed his hair. It flew about 300 meters, it frightened a 30-year-old male passer-by, then floated into a house via a window.

Kornel Toth, a 16-year-old student, noticed the phenomenon and described it as a yellow sphere 20 cm in diameter with a buzzing sound. The intensity of the light was similar to a 100 watt electric bulb. It floated away through another window pushing down the head of József Pasztor, a 33-year-old construction worker, standing under the window. He felt a strong pushing force downwards and saw the yellow light and became frightened. After some seconds everybody in the neighbourhood heard a loud explosion from the house of Vince Adam, some 50 meters away. The owners were at the market, they arrived about 5 minutes after the explosion.

Some of the furniture in two rooms was set on fire. The roof of the building was not damaged, but the 2.5 mm thick aluminium wires in the four room apartment were melted in the walls, and the metal lids of electric function boxes were torn away from the wall ripping off the wallpaper. Several electric bulbs burst, chandeliers were torn off from the ceiling, knobs from the TV set jumped off. The glass neck of some bottled fruit was partially melted, but the fruit inside was not damaged. The estimated minimum energy density of the BL is at least 350 Joule/cm^3 if the melting of a 100 meter long wire is considered alone. The actual length of wire in the walls in the four rooms is between 100 and 200 meters.

3. The following event was recorded in the village of Bekecs, on the 2nd of August at about 12:30 in good, sunny weather (case 384).

Through a window Mrs. Csik saw a reddish ball 30-40 cm in diameter floating horizontally along their yard. The ball knocked out an approximately 10 kg stone from the base of the fence and pushed over all the grape poles (see Fig. 1a). Mr.

Csik, resting on a chair near the grapes, was suddenly lifted off and thrown about 3-4 meters away, but he was not injured. He started to yell for help and Mrs. Csik ran to the yard. She saw that the red tiles of the neighbour's house were moving up and down as if they were being sifted and some of them started to fall (see Fig. 1b).

Suddenly they heard a loud crash and dust billowed from the neighbour's house. They called for the firemen, but there was no fire. After the dust had settled, they saw that the tiles on the roof were extensively damaged and a 20cm x 20cm rafter was broken. The phone cable evaporated and practically exploded from the wall of the hall, where it run inside the wall. All the 1.5 mm thick aluminium electric wires melted in the house (about 100 m), some of the electric bulbs exploded in different rooms. The knobs of a TV set jumped out and some circuits were damaged. All the lids of the electric junction boxes were torn off, and the electric meters broke down. A picture made of metal was thrown to the opposite wall, the metal parts of a barometer jumped off its wooden casing. The closed door of the bathroom was opened by tearing off part of the door post, and above the door the 20 cm glass cover of an electric watch was opened. Near this door, a 30 cm soot spot was found on the wall, but nowhere else in the apartment. Some 5-10 minutes after the events storm clouds appeared, and a short but severe thunderstorm broke out. The estimated minimal energy density of the BL is about 15 Joule/cm^3 for the 40 cm ball, taking into account the melting of at least 100 meters of the electric and phone wires, ignoring the effect of mechanical damages.

4. Next day on the 3rd of August in Fertőd at about 5 a.m., Mrs. Mark saw a white, 30 cm bright luminous ball at the yard of her house (case 419). It was still there about two minutes later, but it moved towards the house of Mr. L. Horvath, some 300 meters away. At the same time, Mr. Horvath woke up in his house, and saw that the weather was cloudy and there was a quiet rain. In the distance he saw a spot of light, like the headlight of a car, at the fields. He went to the bathroom and returned to the bedroom when he and his wife heard a faint buzzing noise which became stronger and louder. In about 4 seconds there was

a moderate crash and the walls started to shake. They ran to their yard and saw dust and fallen bricks all around. When the dust settled, they found the following damage to their house: The 12 cm thick brick wall between the small room and the bathroom was pushed to the opposite wall (see Fig. 2a). The 12 cm thick brick wall between the bathroom and the kitchen moved to the opposite wall, pushing in front of it the kitchen wardrobe (full of chinaware and metal pots and pans, and the furniture) like a piston (see Fig. 2b). The wall between the bathroom and the small room moved in a similar manner (see Fig. 2c).

The upper part of the 40 cm thick brick main wall was pushed out about 10 cm. The window of the bathroom was not broken, it remained intact. The sides of the washing machine standding at the wall opened up. It remained in its original place, there was no damage to its walls. It was not plugged in (see Fig. 2d).

Most probably the phenomenon flew to the bathroom through the chimney because its doors were opened. The gas heated water heater was torn apart in the same manner as the washing machine. No trace of burning was found in any part of the house, and a gas explosion as a possible cause was ruled out by the gas works and the insurance company. There was no damage in the electric system. It must be noted that in the bathroom there were four long 1 inch grounded steel water pipes (bathtub, toilet, kitchen pipes, basin) running outside the walls, so a high current could be conducted away without any special trace.

There were several other observations where the damage was less extensive but which might be of interest from the viewpoint of the physics of the phenomenon.

The common feature of the second, third and fourth observations is that the damaged houses have central heating made of extensive steel tube network, and the houses were located where plain regions meet foothills.

Mr. Z. Unyatinszki (meteorologist) reported the following observation:

"I was working in the Meteorological Observatory at Kékestető (1014 meters above sea level) on the 26th of July. Shortly after 10:30 a.m. the observatory was immersed in clouds, then a thunderstorm broke out. At 11 a.m. we were about to close the windows when we saw a falling, yellowish-grey, shining sphere about 20 meter above ground level moving with the speed of free fall, but not entirely vertically. It was about 1.5 meters in diameter and was slightly transparent. It exploded with a loud bang near, but not at, a 1.5 meter high iron pole at the gate of the observatory. Huge sparks jumped off the iron fence at several points. The meteorological registrating devices were not damaged, but there were disorders at the electric network of the observatory. The power generator switched on, though there was current in the network. At least 10 automatic fuses switched off in an entirely illogical manner: several "slow"fuses switched off, while the fast fuses didnot. The telephones went dead. At the nearby radio station 200 meters away all the shortwave and UHF transmitters broke down. Huge sparks jumped off the metal ceiling of the TV transmitter some 500 meters high.

At the observatory a twin power supply burned down, but its fuses remained intact. In additional to the two of us at the observatory, several people observed the phenomenon from the nearby sanatorium." (case 392).

In four cases the spontaneous formation of the phenomenon was observed without previous streak lightning as in the following case reported by Mrs. Aranyosi.

a) "We were on our summer holiday at Lake Balaton in July. One night we were on the shore of the lake with 11 friends of mine. The water was calm, there was no wind, the stars were very clearly visible and the weather was hot.

Suddenly we saw a yellow sphere emerge from the lake about 100-200 meters away. It had a definite contour, it seemed to be as large as a soccer ball. It ascended at an angle, then vertically for a while, then again at an angle. We watched it for at least 5 minuted then it disappeared. Some hours later a severe thunderstorm broke out." (case 390).

The other three similar cases are also worth describing briefly.

Case 373:

b) On the 15th of July at 7 a.m., five-six 10 cm diameter

whitish balls appeared 2-3 meters away from the 22-year-old lifeguard on the beach of Lake Balaton amidst a buzzing sound. In a couple of seconds they united into a 25-30 cm reddish ball accompanied by a moderate bang. Then the lifeguard ran away. The gardener of the beach saw the ball float to an aluminium covered icecream pavilon (the rest of the buildings were made of brick) 50 meters away, then an explosion was heard. A grounded aluminium electric box (15 x 25 x 30) was virtually welded to the corrugated aluminium wall of the pavilon. The 7-8 mm thick aluminium front piece of the box was torn off, though it was attached with four 8 mm steel screws. Three of them were practically shot 1 cm deep into a brick wall about 150 cm in front of the box, the fourth screw remained in its place but the corner of the front piece was broken. All the electric meters in the pavilon broke down, as did the fuses of the freezers.

The lifeguard complained of sustained sleepiness and pain in the joints.

The weather was calm, though remote thunder was heard from behind the hills some 15 kms away."

Case 394:

c) On a rainy morning of the 26th of July 1987, the wall cracked in the kitchen of Mrs. Kálmán Gál. A long, reddish light appeared which soon became spherical, floated to the gas stove, blew out the flames and disappeared (see Fig. 3a). The telephone cables running on the surface of the wall of the hall melted, but there was no electric damage in the kitchen.

The approximately 5 cm deep crack appeared in the brick wall where the electric cables ran in the wall in metal covered insulated tubing. The tubing was squeezed, crushed at the formation site, but the wires inside were intact (see Fig. 3b). Outside the wall at the same level the pipe one inch in diameter was used for central heating. It was not damaged.

Case 401:

d) On the 24th of July 1987 at about 8 p.m. several 20 cm balls of fire appeared in rapid succession slightly above the ground of a garden 30 meters from the observers. They were yellowish and emitted a crackling sound, then disappeared suddenly and quietly appeared at the height of about 20 meters. There had been some very distant lightning in the preceding minutes.

This "spontaneous" formation is described in other collections as weel[2,3,4] and similar features were reported also in earlier years in the Hungarian collection of observations. This means of formation should be taken into consideration when a theoretical model is proposed.

Figure Captions

Fig. 1a General layout of the damaged house
(1)stone thrown out of the stone fence, (2)row of grape poles pushed to the ground, (3)chair where Mr. Csik sat, (4)electric meters, (5)metal picture on the wall, (6)barometer, (7)sooth on the wall, (8)door torned open

Fig. 1b Tiles of the damaged house

Fig. 2a Part of the general layout of the damaged house
(1)40 cm thick main brick wall, (2)completely ruined wall between the kitchen and the bathroom, (3)almost completely damaged wall between the bathroom and the room, (4)partially damaged wall, (5)washing machine

Fig. 2b View of the bathroom from the kitchen, the place of the wall is visible. ((2) on Fig. 2a)

Fig. 2c View of the small room from the bathroom. Part of the remaining wall, (3)on Fig. 2a.

Fig. 2d The damaged washing machine, its side plates being open, (5)on Fig. 2a.

Fig. 3a View of the kitchen where the ball lightning appeared. It formed at the left-hand side near to the ceiling at the tubes of the central heating.

Fig. 3b The damaged wall with the crushed electric cable tube inside the wall.

References

1) Egely, G.: Hungarian Ball Lightning Observations
Central Research Institute for Physics, Budapest, Hungary KFKI-1987-10/d, 1987.

2) Brand, W.: Der Kugelblitz
Henri Grand, Hamburg 1923.

3) Stahanov, I.P.: Physical Properties of Ball Lightning (in Russian), Energoatomizdat, Moscow 1985.

4) McNally: Preliminary Report on Ball Lightning
Oak Ridge Nat. Lab. ORNL-3938, 1966.

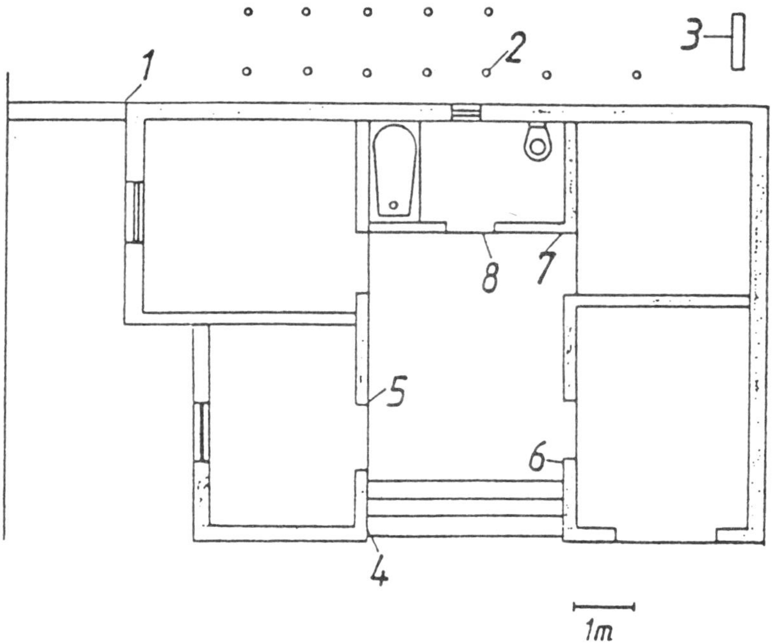

Fig. 1a

Fig. 1b

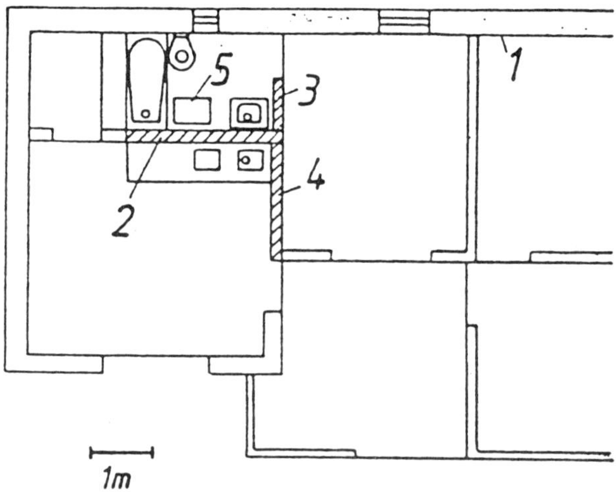

Fig. 2a

Fig. 2b

Fig. 2c

Fig. 2d

Fig. 3a

Fig. 3b

Nature of Ball Lightning in Japan

Y.H.Ohtsuki

Department of Physics, Waseda University
Okubo 3, Shinjuku-ku, Tokyo 169, Japan

and

H.Ofuruton

Tokyo Metropolitan College of Aeronautical Engineering
53-1, Minamisenju 8, Arakawa-ku
Tokyo 116, Japan

We have started to obtain the data of eyewitness of ball lightning in Japan since June, 1986 by the "Japanese Information Center of the Ball Lightning (fire ball)". We succeeded to get 2060 obserbations from all the Japanese islands. (Examples of reports are listed in the Appendix with the Video montage photographs.)

By the detailed analysis of one year, we could conclude that the Japanese ball lightning differs from the continental one reported in Europe[1)] and the United States[2)]. Almost all ball lightnings appear in clear or cloudy, but a very few in thunder. Nevertheless the Japanese ball lightning has a strong correlation with thunder in "space" and "time". The distributions of thunders and the ball lightnings in area and month coincide very closely. This shows the typical properties of ball lightning in this southern and humid island.

Brand[1)] showed that 95% of ball lightnings in Europe is observed in thunder. However, this is not the case in Japan. Figure 1 indicates the observation probability of ball lightning in Japan for different weather. Only 3% of them were observed in thunder.

However, it is very interesting to see the strong correlation between the monthly distributions of thunders and ball lightnings in Japan, indicated in Fig.2. The two curves coincide very closely.

Someone says that this strong correlation comes from the fact that people have a greater chance to be in outdoor in summer season than winter. But this saying is denied by taking the correlation between the area distribution of the thunders and of the ball lightnings.

Figure 3(a) shows the observation ratio (number of observations divided by the population) for each prefecture in Japan. There is so much difference of population between the ocean part (Tokyo, Osaka etc.) and the opposite part (Niigata, Kanazawa) or the northern part (Hokkaido). Thus it is reasonable that the larger number of information on the observation of ball lightning comes from the ocean part. Therefore, it was necessary to divide the number of observations by population.

The prefectures with the hugest observation ratio are marked in black in Fig.3(a), and those with the smallest one in white. The shadowed prefectures in the map of Fig.3(a) show the intermediate area.

It is noted that the black prefectures are very famous for thunders. Figure 3(b) shows the distribution of averaged cases of thunders in August in Japan. There is a precise correspondence between the ball lightning observation ratio and the number of thunder cases.

There is not so much difference in size, color, shape, lifetime and sound of ball lightnings between the Japanese and the continental one[3).

It is not known why ball lightning in Japan is created in clear or cloudy weather, but very rare in thunder. Perhaps humidity will be very important. In the summer season humidity in the Japanese islands increases over 80% even in clear weather. In the continent, such humidity is observed in thunder.

Very frequently, people observed the coincidence of ball lightning (fire ball) with human's death in Japan. We are now analysing these "special kind of the fire ball", comparing also with the thunder cases.

Acknoledgement

The authors thank Prof. S.Kitagawa and all the members of the Japanese Information Center of the Ball Lightning (fire ball) for their helps and discussions.

References

1) W.Brand: "Der Kugelblitz", Hamburg Verlag von Henri Grand (1923)

2) J.R.McNalley: American Division of Plasma Physics, Gatliburg (1960)

3) S.Singer: "Ball Lightning", in "Lightning", Academic Press (1977) p.409

J.D.Barry: "Ball Lightning and Bead Lightning", Plenum (1980)

Appendix

Eleven examples of reports obtained by "the Japanese Information Center of the Ball Lightning" with the photographs from the Video montage. The Video montage were produced by the computer graphic plus video tape of the actual spot acc ng to the eyewitness.

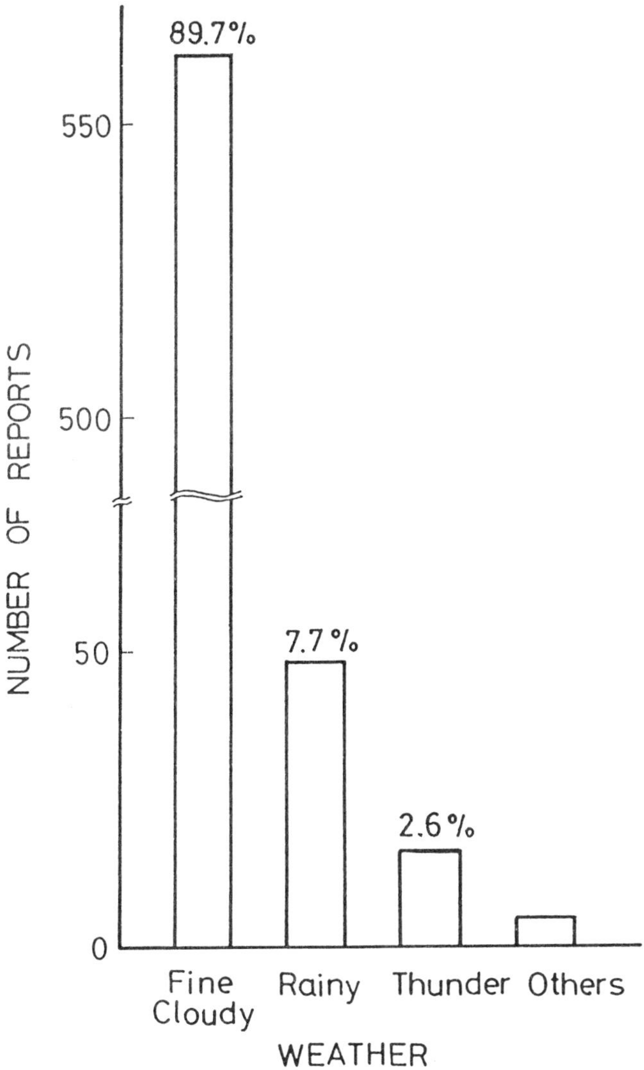

Fig.1 Ratio of the observation reports of ball lightning in Japan for different weather. It is very rare to see ball lightning in thunder, contrary to the continental ball lightning.

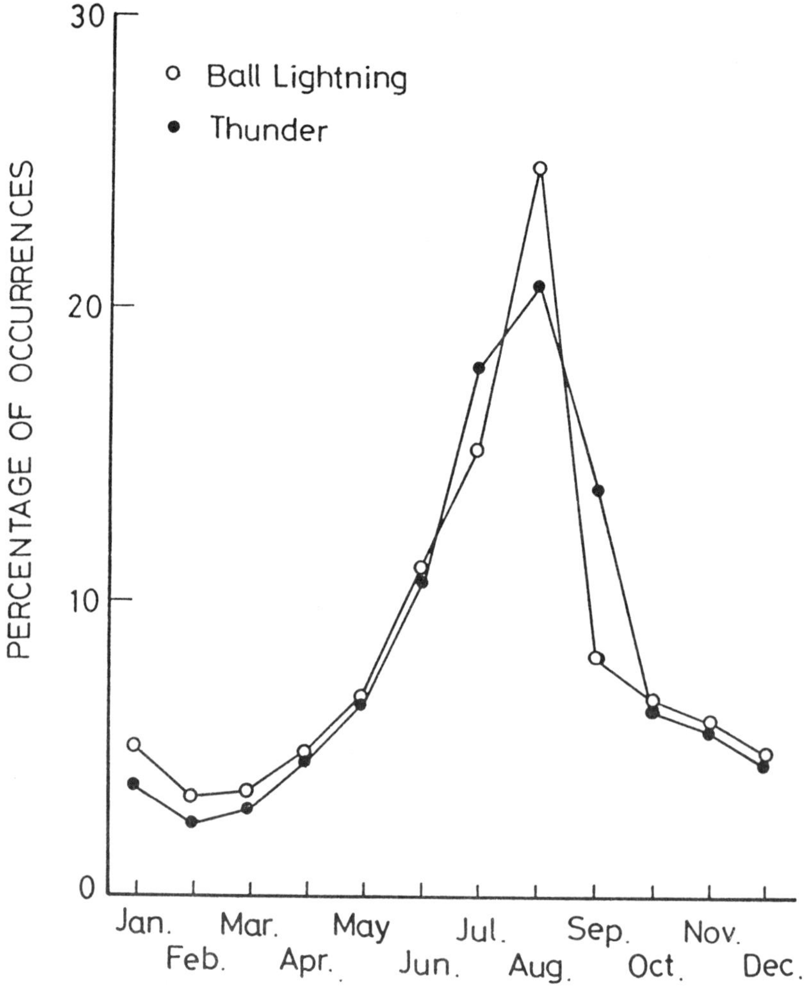

Fig.2 The monthly distribution of thunders (●) and the ball lightnings (○) in Japan, showing the strong correlation between the two curves.

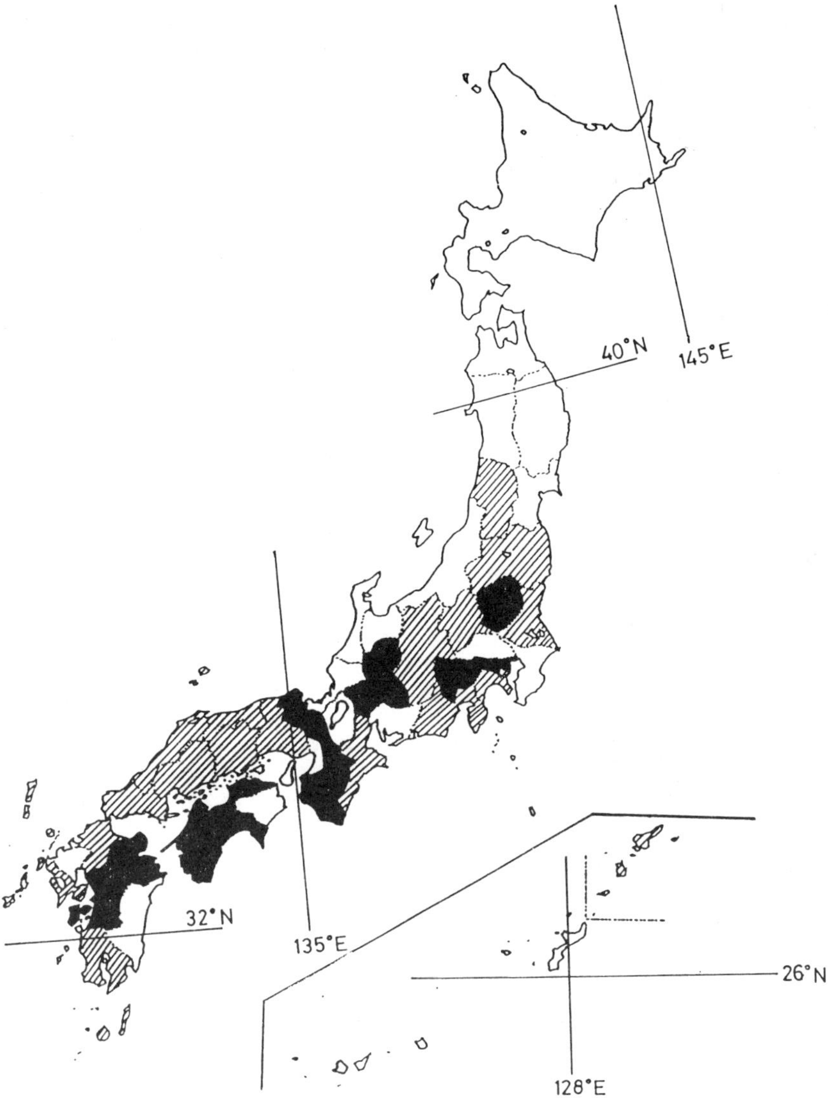

Fig.3(a) The observation ratio (number of observations divided by the population) of ball lightning for each prefecture in Japan. The prefecture with the hugest observation ratio are marked in black, and those with the smallest one in white.

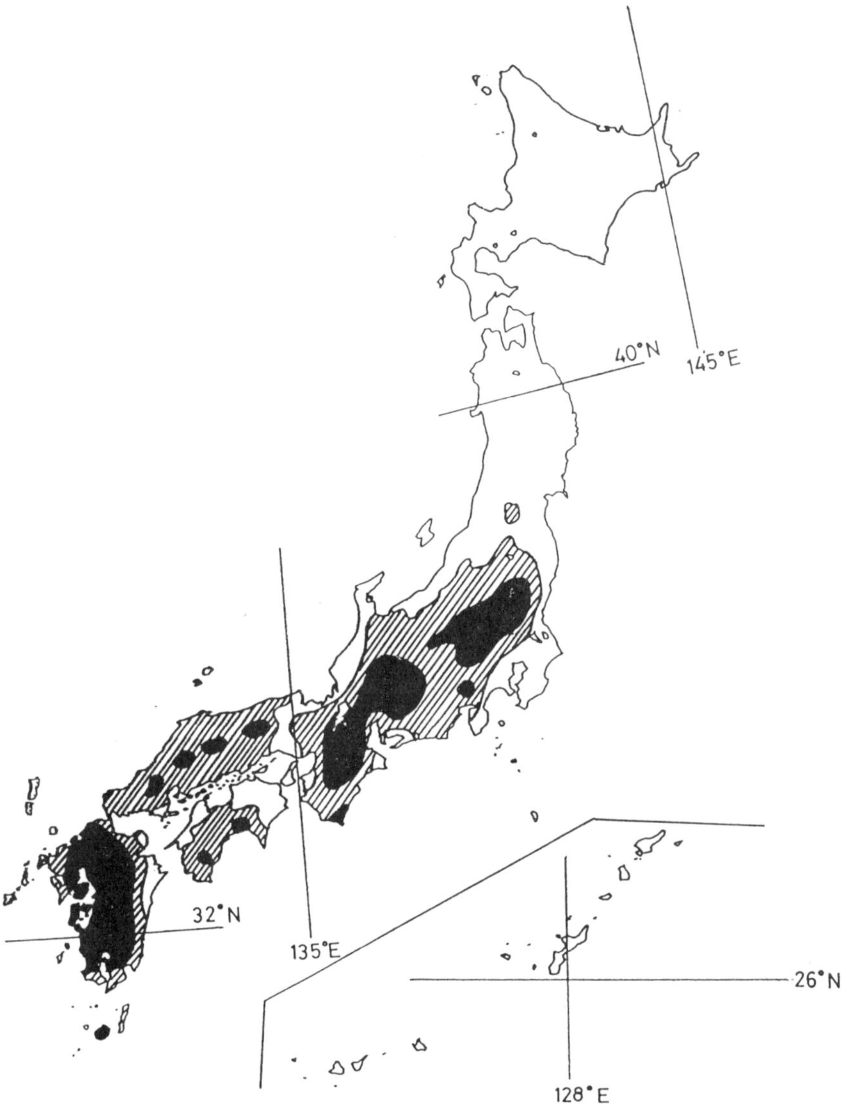

Fig.3(b) The distribution of thunders in the Japanese islands.

1.

Mr. S.S. (Tokyo) (32 years old)

It was one clear evening of summer(8:30 p.m. in June 1980). I was waiting for a train at Kichijoji Station (Tokyo). I saw an orange color lighting ball rising in the distance. It was brighter than helicopter lamps. After 5 or 6 seconds, it changed the course and accelerated. It was flying across the sky for 2 or 3 seconds and it disappeared without sounds.

1-1

1-2

1-3

2

Mr. T.T. (Tokyo) (? years old)

At 9:20 - 9:23 p.m. on July 21th 1987 the weather was cloudy. When I was on my way back home and I turned a corner, I saw a lighting matter floating 1 m above the road. I neared it, but it went away from me. I ran after it for 1 min, but I could not cut down the distance. It broke and disappeared at the one side of the road as if a balloon had bursted.

The shape of the lighting matter was sphere with a small tail. The color of its center was orange and red, and its tail was blue. The diameter was 25 - 30 cm.

2-1

2-2

2-3

2-4

2-5

2-6

2-7

3

Ms. Y.Sano. (Kanagawa) (? yaers old)

In july of the end of 1970s, it was a fine day and the temperature was 27 - 28 °C. At about 7:00 p.m. I was walking up the sloping path at Den'enchofu (Tokyo) with a friend of mine and we saw a red ball moving slowly and horizontally in the trees. Its diameter was 20 - 30 cm. The whole phenomenon lasted about 10 seconds, then it disappeared without sounds.

3

4 Mr. M.Yamamuro (Tokyo) (74 years old)

In June of 1927, I was 13 years old. It was one clear evening (7:30 p.m.). I was standing on a road near the railway bridge of Ebisu Station (Tokyo). I saw a ball lightning with a long tail floating and moving to the railway bridge.

The color of its center was orange, its surrounndings was light yellow, and its tail was light blue. The diameter was 30 cm and the length of the tail was 3 - 4 m.

I ran after it until it disappeared. It was moving over the railway bridge and disappeared in the trees near the railway bridge.

4-1

4-2

5 Mr. A.Hayashi (Tokyo, 74 years old)

It was one clear night of winter (10:05 p.m. on January 30th 1987). When I was walking at Todoroki (Tokyo) on my way home, I saw a ball lightning moving slowly and horizontally. There were electric wires near the ball lightning and the height of them was about 5 meters. The ball lightning was moving on this side of the wires and along them.

The color of its center was red and its surroundings was yellow. Its shape was sphere with a few small tails. The whole phenomenon lasted about 2 seconds. I could not hear any sound.

5

6 Mr. H. Ofuruton (TOKYO, 29 yaers old)

In the summer of the end of 1960s in Tokyo, I saw a small red ball lightning floating and moving out of the window. I could see it for about 2 seconds, and it went out of my sight. Its diameter was about 3 cm. I could not haer any sound. The weather was fine.

6

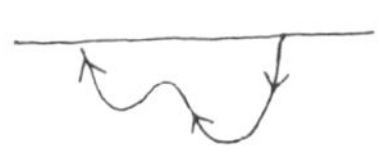

7 Mr. K. WADA (MAEBASHI, 43 years old)

On 30th August 1987 my wife and I went to Kirigamine highland (Nagano pref.) to take photographs of the view of dawn. The weather was fine without any cloud. Stars were twinkling.

At 4:20 - 5:00 a.m. we walked on a path from a hut for about 20 minutes. When we turned around to see the hut, suddenly an orange color light began to shine in the west sky. The light became brighter and brighter, and reached the strongest brightness. After 10 seconds, the light became weaker and weaker, turned white color and disappeared.

In a minute it appeared in the white color. It became brighter and brighter, turned to orange color and reached the strongest brightness. It lasted for 10 - 20 seconds. The light became weaker and weaker, and disappeared.

My wife told me that there was another light in the righthand side, and I saw the light there. She told me that it appeared many times, but I could see it once. Moreover she told me it also changed the color, but I did not see such a phenomenon.

7

8 Mr. M. SHIBATA (NAGOYA, ? years old)
Mrs. M. SHIBATA (NAGOYA, ? years old)

In about 1980, we do not remember the season, the weather of the day was ordinary, not rainy and not windy. At 8:00 p.m. we saw a orange ball lightning floating and moving just over the roof of the neighbor's house.

Its shape was sphere larger than a pingpong ball and smaller than a baseball. It was moving horizontally for about 3 seconds. We could not hear any sound.

8-1

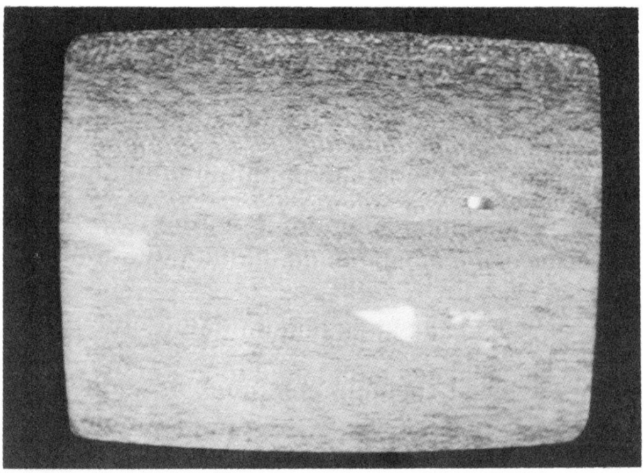

8-2

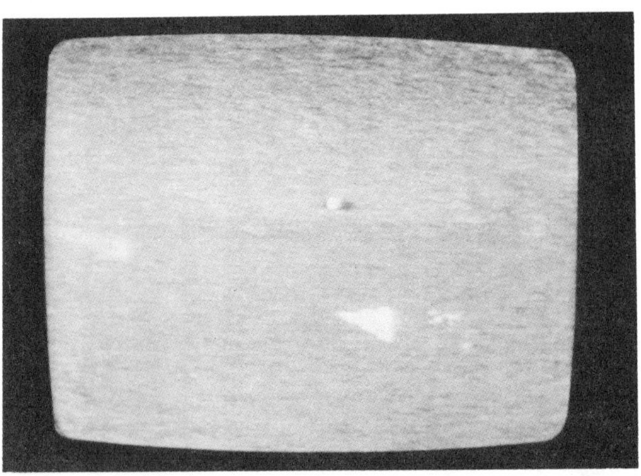

9 Mrs. H.Sato (Saitama) (? yaers old)

In the summer of about 1982, our family was walking on the bank of the Arakawa River (Tokyo). I do not remember the weather. At about 8:00 p.m. a fire ball appeared in the sky. The color of its center was green or blue and its surroundings was whity. It moved horizontally, then moved down, and disappeared behind a building. I could not hear any sound.

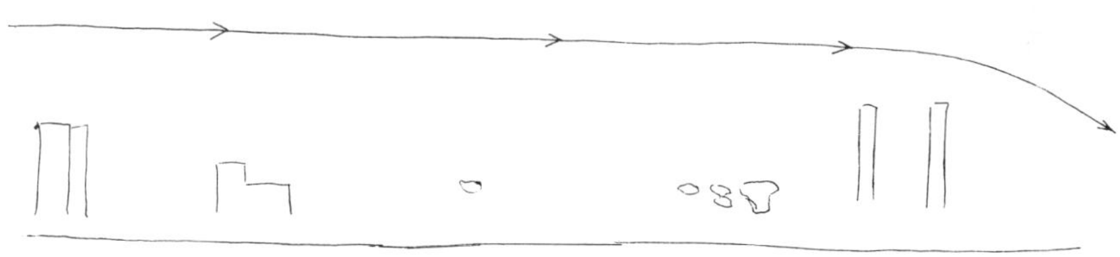

9-1

9-2

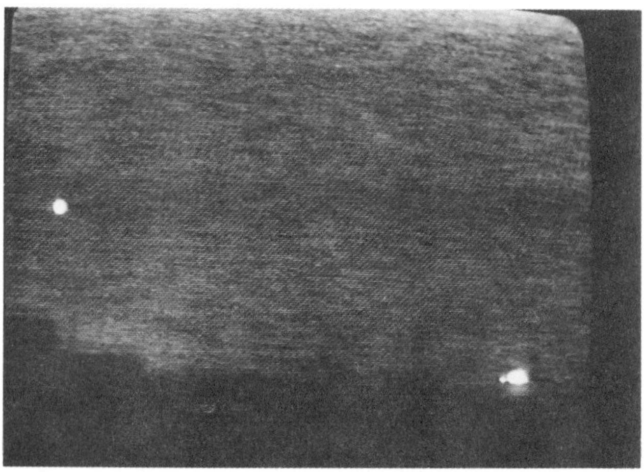

9-3

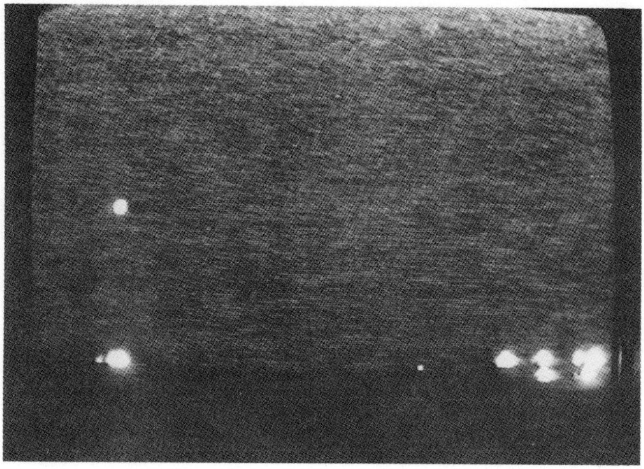

9-4

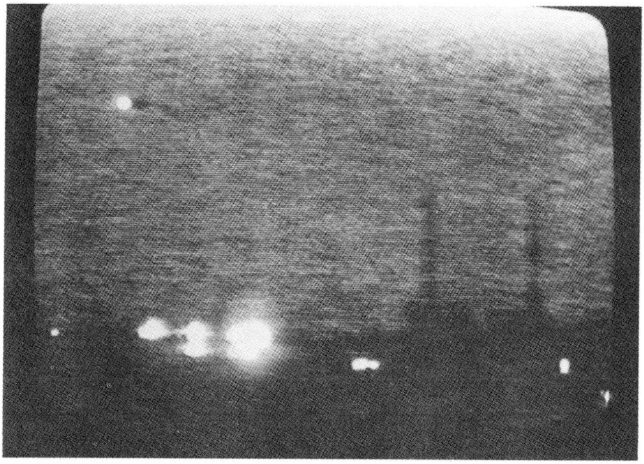

9-5

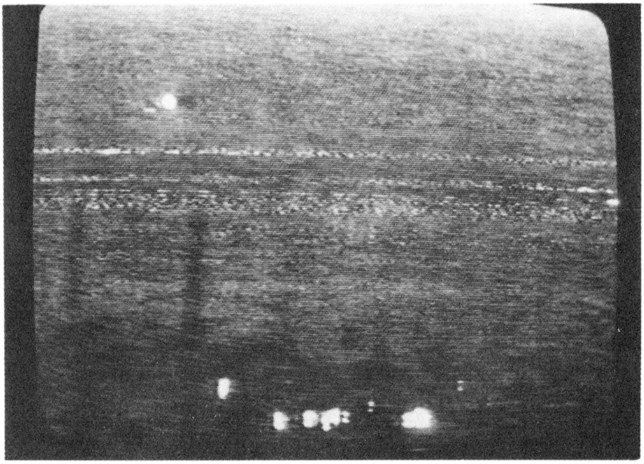

9-6

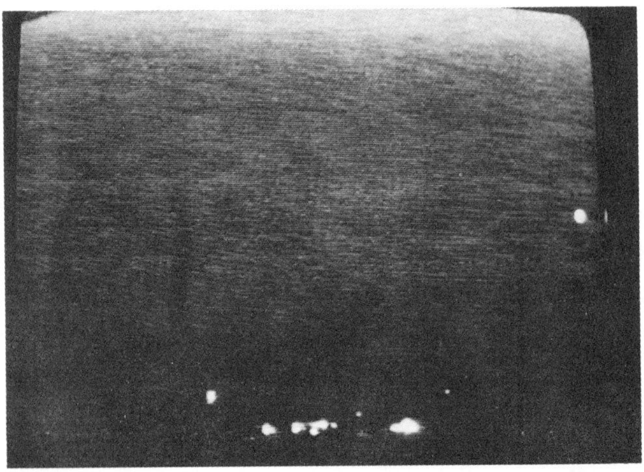

10 Mrs. M. TAKADA (NAGOYA, ? years old)

At about 8:00 p.m. on December 4th 1983, my husband was driving a car and I was sitting on the next seat. The car was running on a expressway in a mountain district between Kyoto and Nagoya.

A red ball lightning was crossing the front glass from right-up to left-down. Its diameter was 30 cm. The whole phenomenon lasted 2 - 3 seconds.

The weather of the day was snowy in the morning and fine in the afternoon.

10-1

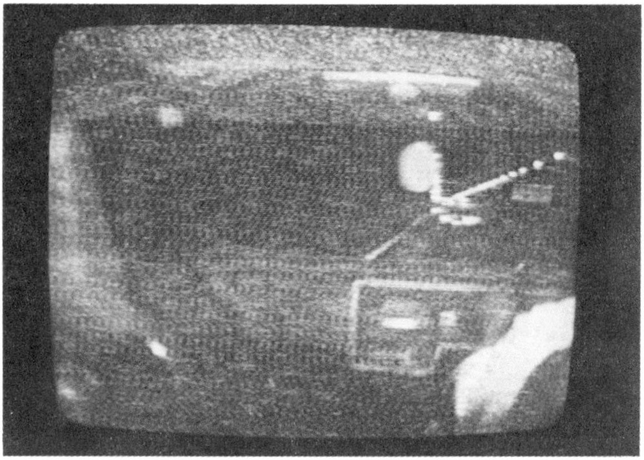

10-2

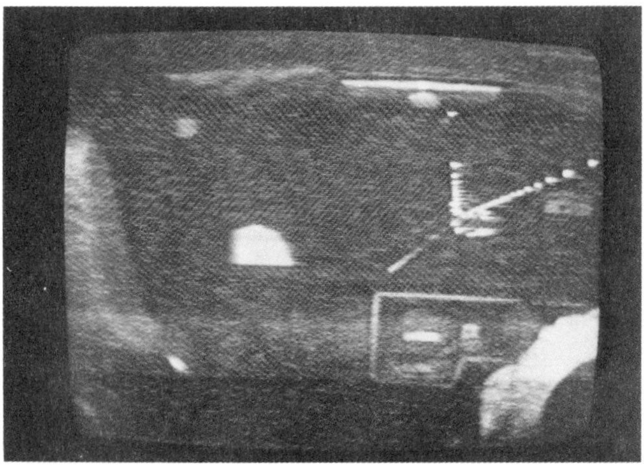

11 Mrs. M.T. (Kanagawa) (? years old)

At about 1:30 a.m. on July 7th 1977, the weather was a thunderstorm. I was sleeping in my bed room. Suddenly I heard a loud noise and a lightning flash fell from the ceiling to me, and I felt a great shock to my arm. The noise, the flash, the shock, they happened simultaneously. At that time, television screen of our house became white and I could smell something burning.

The next morning I found that the antena of the television was broken and that a wire fastening the antena was cut. Moreover the telephone wire was cut, the fuses were blown, and the plug of the television code was broken. The television was beyond repair. A dog out of the house was dead.

(However, it is very difficult to make a video montage of this case. So we cannot present a photograph of the video montage.)

PHENOMENOLOGICAL AND PSYCHOLOGICAL ANALYSIS OF 150 AUSTRIAN BALL LIGHTNING REPORTS

Alexander G. Keul & Kurt Schwarzenbacher, University of Salzburg, Austria

§1 Basic information about Austria and the Ball Lightning (BL) project

Austria is a neutral state in Central Europe between the west and east bloc. Austria has 7.5 million inhabitants on a territory smaller than Florida (USA) or about the size of Hokkaido (Japan). Two thirds of the territory are made up by sparsely populated, high mountains (the Austrian Alps), the remaining one third by densely populated plains and valleys (Figure 1a, 1b). Austria has a moderate climate with prevailing western winds bringing in rain from the Atlantic Ocean. Thunderstorms have a frequency between 15 and 30 days per year, with a maximum area in southeastern Austria (Figure 2).

BL was and still is a subject of folklore in Austria: The "babbling ball" telling the future to good people or the "skopniak", a mischievous fiery dwarf, are old forms (Figure 3). Modern Austrian folklore links any unusual damage or unexplained thunderstorm effect with BL. This dramatized, false picture of BL is spread by the local police and press.

Scientific attention for BL in Austria developed 30 years after Arago (1838): Von Haidinger (1868 a,b,c) and a private natural history society collected and published first reports. 13 of them are quoted in the famous book by Brand (1923). Other BL articles came from Prohaska (1900, 1901, 1903), Trapp (1949), Neunteufl (1951), Pühringer (1965, 1967), Zapletal (1970) and Charlier (1966, 1972-1973). They were mostly descriptive and speculative - "mixed pickles" without systematic attempts.

When in contact with Pühringer in my Vienna days as a meteorology student, I started with BL field investigations (Keul, 1977), made press calls-for-reports and published a simple statistical analysis of 65 Austrian case histories (Keul, 1980, 1981). Ten more press calls in 1981 and 1982 helped to double the number of cases. Prof.Ohtsuki's letter in 1987 was the "spark" that set off a new, computer-based evaluation of 150 reports.

§2 Short illustrative examples of Austrian BL cases

I will give you a few field-investigated examples of Austrian BL cases in a short, narrative form:

-Summer 1971, Vienna city area: A worker looked out of the window during a thunderstorm with rain. He saw a bright globe falling down into the lane, splitting into three identical balls upon impact. Two rolled on in the lane, but the third one passed through a metal garden fence and disappeared in the garden. No damage was done to the fence.

-Spring 1972, Graz city: A pensioned teacher watched TV on a damp, rainy evening. Three other persons in the yard of the same house observed a bluish, hazy ball flying a "wavy" trajectory to the open window of the teacher's flat, where it exploded with a loud bang. A tickling, "electrified" sensation was present in- and outdoors some time after the incident.

-Summer 1976, Graz city: Lightning struck the house of a family, damaging telephone and electrical appliances. Seconds after the clash, a bright, yellow ball "two times a cherry" jumped down the staircase step by step "like a rubber ball", disappearing soundlessly. At the first floor where it originated, a wall damage in the form of a hole was detected.

-Summer 1979, open countryside in southeastern Austria: A farmer saw a blinding ball of light in the evening, near a local power line. When phoning the power company, we heard that the 60 kilovolt line had been interrupted at the time of the observation due to a short-circuit (forming an arc). This explained the report.

-Summer 1979, Vienna city: A pensioned electrician was at home in his flat during a violent thunderstorm with hail and floods. Suddenly, his pet poodle fled in panic under a chair, when a yellow-red, transparent globe appeared near the gas stove and moved towards the observer, ending up with a "deafening bang" and bright rays. No damage and no odor was noticed.

§3 Ideas and method of the 1988 BL evaluation

The main shortcoming of present BL research is the lack of a good working definition for the phenomenon under study. Revers, a late psychology teacher of mine, always emphasized: "**What** do you want to study?" As long as it is unclear, what **is** BL and what **is not** BL, theoretical confusion with hypotheses from afterimages to black holes, and a polarization of researchers between "believers" and "non-believers" is quite natural. Circular definitions like "during thunderstorm.. luminous.. globular" (Barry, 1980, p.33) or "elusive globe which appears in thunderstorms" (Singer, 1980, p.332) do not help very much. When even these redundant pieces of information - round and thunderstorm-association - do not hold in a fair amount of cases (Ohtsuki, 1987), we are approaching the unfortunate UFO "definition", where anything inexplicable automatically becomes a member of this "dustbin category". Reports of St.Elmo's Fire, corona discharges, bright meteors, short-circuit arcs, will-o-the-wisps, earthquake lights, glow-worms and fireworks are **not** excluded by an open BL definition.

Our evaluation idea was not to use a narrow BL definition and throw out aberrant cases, but to take the risk of including marginal phenomena like St.Elmo's Fire seen as a ball, and then to cluster cases according to their phenomenology to see what is the Austrian "core phenomenon" and what is a "borderline case".

Until 1988, we had collected 250 raw reports. 45 were excluded from this evaluation because they were explainable lightning or high tension effects like the one in the previous case example. 55 more were excluded because of insufficient information, for instance short old press reports. 150 case histories remained. Of these, 16 have been field-investigated personally, 110 observers were interviewed on the phone or sent a questionnaire after their first reports, and 24 written reports were not directly investigated by us.

The coding was done by the authors (who cross-checked each other) and used a total of 28 qualitative and 14 quantitative variables. The observer variables were sex, age, profession, reported eyesight, reported reaction, and number of witnesses. The situation and environment variables were date, local time, location pixel, sea level, distance, observer geography and the weather. The phenomenon variables were number of objects, duration, object geography, height, motion, object relation (between object and other things), shape, size, outline, color, brightness, odor, sound, sparks, tail, residue, rotation, size variation, formation and disappearance.

The final data field was evaluated at the computing center of the natural sciences faculty at Salzburg University using the BMDP statistical software package, program version 1987.

Results of the 1988 BL evaluation - phenomenology and psychological effects

The results of our evaluation were the following (Table 1):

The witnesses showed the same sex distribution as the Austrian population. The mean reporting age was much higher than the mean observation age. This seems to be a natural selection effect - when asked to report, chances for old people to have seen BL during their lifetime are higher, so the reporter population is over-aged, but the sighter population is fairly normal. The number of reporting academics was higher than in the general population, the self-reported eyesight seemed to be better than on the Austrian average. Every second case had a single observer, but there were up to ten witnesses in some events.

Five years (1950, 1963, 1972, 1976 and 1977) showed more cases than the rest. There was a distinct summer peak coinciding with the Austrian thunderstorm maximum (July). The-time-of-day peak again overlaps the frequency-of-lightning-discharges maximum (3:30 - 4:30 P.M. Central European Time). For the sources of the meteorological statistics, see Keul (1981). Most cases came from Vienna and Styria (the southeastern Federal Province), where the population maximum (Vienna) and the thunderstorm maximum (Styria) are located (Figure 4). Report frequency falls off with rising sea level, but there still are cases at heights of 3000 meters and more. The great majority of events happened within settlements and the objects were observed from indoors, which is not unusual under thunderstorm conditions. Every second object was as near as five meters or less to the observer, and 75% moved within a 100 meter radius. BL is closely linked with thunderstorms in Austria (72%), but not so close with precipitation (mostly rain). A near lightning stroke was mentioned in about 40% of the reports. Sometimes, static electricity was present.

The objects were nearly always single phenomena with up to ten seconds duration. 80% were seen inside settlements, two thirds of them outdoors, one third happened indoors. Ground level was the typical height, and the characteristic motion was horizontal or downwards. Some showed a hopping or vibrating secondary motion. Every second object had physical contact, i.e. touched a bad conductor (like the floor, walls) when moving, one third was related to good conductors (electric cables, water), the rest floated freely. BL is a correct term in Austria - objects are always round to oval. The diameter distribution has a mean of 30 centimeters, but every second object is smaller than 20 centimeters. The outline is sharp in most cases. The red-orange-yellow sector is the predominant color, but there are also clusters of blue-violet and white objects. The overall brightness is not blinding and steady. "Special effects" noticed in some cases are odors, simultaneous sound, sparks, a tail, rotation and size variations. Formation of BL is quite frequently seen (25%). About 40% of the objects ended with a terminal explosion, usually a loud bang, but nearly the same number disappeared without sound. "Residue" reported after BL events could be identified as normal lightning damage in most cases, but there were also cases of BL starting a fire or scorching things.

The overall phenomenology is not significantly different from the picture obtained by the first evaluation done in 1980 (Keul, 1980, 1981). As an expansion of our files from 65 to 150 cases has not upset the patterns, we assume that 100 or 200 additional cases from Austria would not be too different, either. The "core phenomenon" of BL in Central Europe seems to be stable enough to use a working definition for it.

So far, we have only looked at isolated variables and their distributions - an elementaristic approach. It neglects the natural links between variables. To detect more structures, we used a block clustering program for categorial data (BMDP3M) for nine variables: Sea level, thunderstorm, precipitation, lightning stroke, duration, primary motion, object relation, diameter and color. In a successful block clustering, the total number of block counts should be a high proportion of the total number of categorial data values. In our case, it was only 58%.

As a more promising way of evaluation, we tried a correlation analysis. 20 variables (13 qualitative and 7 quantitative) were used in BMDP4F for two-way frequency tables, Spearman rank correlation coefficients and the equivalent t-test values. We computed a total of 200 correlations and obtained 16 Spearman rank correlation coefficients with highly significant t-test values (according to the alpha-adjustment for 200 correlations, the 1%-level is still significant, here). In this computer run, "delta", the difference between observation and reporting age of the witness, was correlated with all salient BL variables to see whether older and older cases show a drift in reported phenomenology as suspected by Charman (1979, p.281-282). The program detected two correlations (delta and number of persons, delta and emotional reactions), one nonsense, one psychological, but no drift of phenomenal details. The "tree" of significant correlations shows a positive one between diameter and distance, a negative one between distance and reported emotional reaction, and so on (Table 2). Interpreting the non-trivial correlations, we notice:

- Reported diameter increases with reported distance (I)
- Reported fear reactions decrease with distance and elapsed time between observation and report (II)

Weaker correlations are:

- The number of persons increases with elapsed time and sea level (nonsense correlation?)
- Reported diameters are bigger without lightning strokes
- Secondary motion (i.e. hopping) decreases with reported diameters
- Objects associated with good conductors are thunderstorm-bound
- Primary motion increases with reported duration

Of the two strongest results, II is easily understood. Fear correlates negatively with distance, and the reason why fear reactions are less reported with increasing time distance between event and report was called "repression" (of unwelcome emotions) by Sigmund Freud.

Effect I is not so simple (Figure 5). There is no physical reason why more distant objects should be systematically bigger. As Altschuler (1969, p.730-731) has already mentioned in his summary on BL, a positive distance-diameter correlation would either mean that more distant objects are overestimated because of a lack of reference objects (which is unlikely), or that distant objects would have to be bigger to attract attention and smaller ones pass unnoticed (which is more plausible). Irradiation (greater subjective size of bright objects) is dependent on object brightness and the adaptation level of the eye, so it cannot explain the distance effect. As the median distance object-observer is two meters indoors and 55 meters outdoors (Altschuler: 3 versus 30 meters; 1969, p.730), and the median BL diameter is 12 centimeters (mean: 23 cm) indoors and 25 centimeters (mean: 30 cm) outdoors, I suggest for the future to separate "BL I" (indoors or less/equal 10 meters distance) and "BL II" (outdoors or more than 10 meters distant), or two categories of a better definition, instead of mixing the two distance categories beyond statistical recognition.

The weaker correlations look like genuine physical information at first sight, but need to be confirmed by other studies to be meaningful.

I will sketch you six special evaluations that were made to answer questions which came up in course of the general evaluation:

Do BL events really happen at random? An approximation of the Binomial distribution -the Poisson distribution- can be used for the statistical representation of stochastic, rare events. Misprints per page of a given book, or deadly accidents per year and given mile of a highway are poisson-distributed, if random. Austrian lightning stroke frequency per square kilometer is about three per year. Together with the estimated BL frequency of 0.3% of all lightning strokes (Barry, 1979, p.308), this gives a maximum observation probability of one every 10 years. Counting the number of BL events per year and given Austrian pixel, we got values that are correctly approximated by a Poisson distribution except for the observed values of two pixels (Table 3a). The Vienna and Styria pixel produced series of BL reports falling out of a Poisson distribution. We conclude that BL observations happen in a random way except when population density and/or thunderstorm frequency is high. Also, the number of indirect lightning-strokes via the grid in densely populated areas is considerably higher (Gugenbauer, 1988), and causes additional BL.

How good are BL duration estimates? It is problematic to calculate a simple arithmetic mean of BL observation duration values. First, the observation may cover the total duration of the BL phenomenon, or only a part of it. Second, not all observers have the same expertise to estimate the subjective duration of the event. It is known from experimental psychology (e.g.Funke, 1988) that reconstruction of an event ("re-living" it in the mind) gives more realistic duration estimates than simple guessing. In the Austrian file, we have all types of complete or incomplete observations (Figure 6). There are cases where the witness just tried to guess the duration, and field-investigated cases where a duration reconstruction was done. Evaluating 109 Austrian cases, I noticed that although it is not possible to interpret the results in a strictly quantitative way because the visible path of BL is different from case to case, there is a trend of reconstructed durations to be shorter than guessed ones, and of the standard deviations to be smaller (Table 3b). Although anxiety and unpleasantness seem to lengthen duration estimates (Fraisse, 1984, p.23-25), there was no significant correlation between reported emotional reaction and duration estimates. Interpretation is difficult, here - Funke (1988) detected high interindividual differences, but Drake (1974) found a good matching of witness duration data on fireballs with instrumental data. In any case, I suggest that a duration reconstruction is done with the witness at the location of his observation whenever possible to obtain more realistic subjective time data. In case you are still not convinced that a very short emotionally strong event may be blown up to a minute when guessing, just go and visit your dentist.

After all, are BL reports highly subjective, imaginary stories? In our file of 150 reports, 19 events were multiple reporter cases (Table 4a): Eight people had reported two BL events each, one person even three events. This offered a chance to find out whether two reports of one observer are very similar (as should be the case when the observation is highly stereotype and subjective) or rather dissimilar. Four reporters (A to D) had seen two events in different locations, four reporters (E to H) had their two (respective three) encounters at the same location. Looking at 18 phenomenon variables, I counted the number of different variables from case to case. A location effect was apparent, but not really strong. What was more interesting: Eight of the nine variables that were different in three cases or more, were central phenomenon variables. Of course, it is not possible to deduce whether varying descriptions of identical phenomena or correct descriptions of varying phenomena were given, but it should be noticed that multiple observers have no pronounced BL stereotype.

Are Austrian BL cases without thunderstorms different? Case reports violating the inofficial BL definition are rare in our Austrian files. Only nine events were counted. Most of them also mentioned no lightning stroke (7), no precipitation (6), and no electrostatic phenomena (5). Durations, distance, diameter and color of the objects scatter considerably. No "typical non-thunderstorm BL" is noticed, but it looks rather like a residual category.

Is it possible to link Austrian BL with St.Elmo's Fire? According to Elster and Geitel (c.f.Gockel, 1925, p.79-97), who observed St.Elmo's Fire at the Sonnblick mountaintop observatory in the Austrian Alps, the electrostatic discharge phenomenon is always associated with precipitation, but not always with thunderstorms. A faint sizzling sound is heard, and when daylight is weak, or particularly at night, red, blue and/or white light effects are noticeable. The phenomenon happens as a brush discharge on protruding objects. To check our BL data set for a link with St.Elmo's Fire, I scanned the variables precipitation, object relation, color (red, blue, white) and simultaneous sound (Table 4b): Of the 32 cases where at least two of the four St.Elmo-parameters coincided, 24 or 75% had reported precipitation. Nine precipitation cases reported white objects, seven reported red objects, four reported blue or violet objects. If we take into account that this combination includes 21 or about half of all 44 white, red and blue objects in our Austrian data file, it does not look like a mere coincidence. The borderline between the two phenomena seems to be fuzzy. A similar evaluation should be done for BL and short-circuit phenomena (electric arcs).

Is BL more frequent or more rare on mountaintops? We received two old reports of BL from a (retired) meteorological observer who had worked at the Mt.Sonnblick weather station (3105 meters, Austria's highest observatory) for 34 years. As Prof.Berger, Zürich, had based his anti-BL-article (1973) on the fact that he did never witness or photograph BL during many years of lightning observations on Mt.San Salvatore near Lugano, Switzerland, I thought it would be of interest to investigate BL on mountaintops a little further.

I phoned seven more meteorological observatories on mountaintops (Mt.Schöckl, Feuerkogel, Patscherkofel, Zugspitze, Wendelstein and Hohenpeißenberg) in Austria and Southern Germany, ranging from under 1000 to nearly 3000 meters above sea level and asked the weather observer whether he or a colleague had ever seen BL. The result was negative for all stations - even senior weather observers working for 30 to 40 years on the mountaintops had not witnessed BL. Only a retired weather observer from Southern Germany had seen BL, not on the mountain, but at home in his village. Since only four (2.7%) of our 150 cases happened on mountaintops, this form of BL seems to be a phenomenon with almost zero probability in Central Europe. Therefore, Prof.Berger's reasoning that BL could not exist because he did never see it on his mountaintop, is erroneous.

§5 Comparison with other studies and suggestions for future BL research

What can we learn from this evaluation in comparison to other studies, and are there any suggestions for future projects?

It is the dilemma of BL research that we are mostly studying **reports**, not physical evidence, and that an assessment of the reporter is problematic years after the event. Keeping in mind that the data are of varying quality (compare Charman, 1979, p.280-284) and influenced by a "sociocultural stereotype of BL" (Charman, 1979, p.281-282; Kleiter, 1988), we can try to compare results from different data banks (Table 5) cautiously: You can see that the similarities are greater than the discrepancies. With internationally standardized data collection and evaluation procedures, it would be more easy to interpret the differences.

So, what are the perspectives for future studies? They should try to get a benefit from closer cooperation between laboratory and field researchers in natural and social sciences. The BL report as phenomenon is a very complex affair, and the plasma physicist is rather helpless without psychological expertise - and vice versa. Interdisciplinarity should be more than a word, here. Psychology as a social and natural science of human perception, memory, cognition and emotion has been neglected almost completely in BL research. Two examples: Even basic books on the psychology of perception describe the fact that very bright light sources or objects in dim light are subjectively perceived and reported as "round" (Vernon, 1971, p.49-50). Haines (1980, p.226-227) has successfully tried to capture this effect with a camera. Bright lights that are blinding were shown to produce a rapid shift of subjective colors (experiment: Cornsweet et al., 1958; field: Drake, 1974). Thus, we should rate blinding or high-contrast objects as "low reliability observations", not throwing them together with non-blinding object data. Second, it is known from social psychology that the perception and recollection of visual structures can be influenced by verbal Gestalt cues, i.e. in a "leading" question about them (Carmichael et al., 1932). It is important **not** to ask questions like "How did the ball look like?" in field investigations.

The "naive realism" (believing what the witnesses say) of the old days may give way to a "sceptical realism", i.e. trying to reconstruct what happened before a verbal label is put on the experience by the witness and researcher (e.g. the duration **experience**, not only the guessed duration). A lot could be done, like a BL sighting simulation in the laboratory or other things, but few people will find it rewarding. Sporadic, controversial subjects are not the favorites of scientists under time, financial and success constraints. On an international level, we need at least a **working definition** of what we are studying, and a **standard investigation and reporting system** to have comparable case material from different countries instead of "report salad". Our meeting in Tokyo could be a starting point.

References:

M.D.Altschuler (1969) Atmospheric Electricity and Plasma Interpretations of UFOs, in: E.U.Condon, D.S.Gillmor (Eds.) Scientific study of unidentified flying objects. Bantam, New York, 723.

F.Arago (1838) Sur le tonnerre. Annuaire au roi par le Bureau des longitudes, Notices Scientifiques. France.

J.D.Barry (1979) Frequency of Ball Lightning Reports. Journal of Geophysical Research 84, 308.

J.D.Barry (1980) Ball Lightning and Bead Lightning. Plenum Press, New York.

K.Berger (1973) Kugelblitz und Blitzforschung. Die Naturwissenschaften 60, 485.

W.Brand (1923) Der Kugelblitz. Henri Grand, Hamburg.

L.C.Carmichael, H.P.Hogan, A.A.Walter (1932) An experimental study of the effect of language on the reproduction of visually perceived form. Journal of Experimental Psychology 15, 73.

L.A.Charlier (1966) Kugelblitz-Beobachtung in Vorarlberg. Wetter und Leben 18, 204.

L.A.Charlier (1972-1973) Bericht über Kugelblitzeinschlag. Wetter und Leben 24, 260; 25, 49.

W.N.Charman (1979) Ball Lightning. Physics Reports 54, 261.

T.N.Cornsweet, H.Fowler, R.G.Rabedeau, R.E.Whalen, D.R.Williams (1958) Changes in the Perceived Color of Very Bright Stimuli. Science 128, 898.

E.M.Dewan (1964) Eye witness accounts of Kugelblitz. Air Force Cambridge Research Laboratories CRD-125.

F.D.Drake (1974) On the Abilities and Limitations of Witnesses of UFOs and Similar Phenomena, in: C.Sagan, Th.Page (Eds.) UFO's - a scientific debate. Norton, New York.

P.Fraisse (1984) Perception and Estimation of Time. Annual Review of Psychology 35, 1.

J.Funke (1988) "Changes" oder "Effort"? Eine Überprüfung von zwei Theorien zur Zeitschätzung. Zeitschrift für experimentelle und angewandte Psychologie 35, 218.

A.Gockel (1925) Das Gewitter. Dümmler, Berlin.

A.Gugenbauer (1988) personal communication, Linz, Austria.

R.F.Haines (1980) Observing UFOs. Nelson-Hall, Chicago, Illinois.

A.G.Keul (1977) Kugelblitz in Wien. Wetter und Leben 29, 188.

A.G.Keul (1980) Kugelblitze. Wetter und Leben 32, 167.

A.G.Keul (1981) Ball Lightning Reports. Die Naturwissenschaften 68, 134.

G.D.Kleiter (1988) personal communication, Salzburg, Austria.

C.Lüdecke (1982) Phänomenologie der Kugelblitze. MAN Technologie GmbH, München.

J.R.McNally (1966) Preliminary Report on Ball Lightning. Oak Ridge National Laboratory, Tennessee, ORNL-3938.

J.Neunteufl (1951) Wintergewitter, St.Elmsfeuer und Kugelblitz. Wetter und Leben 3/4, 49.

Y.H.Ohtsuki (1987) letter to Alexander Keul, January 8.

K.Prohaska (1900) Blitzschäden in Steiermark und Kärnten im Jahre 1899. Meteorologische Zeitschrift 17, 331.

K.Prohaska (1901) Feuerkugel in Pöllau, Steiermark. Meteorologische Zeitschrift 18, 94.

K.Prohaska (1903) Über Blitzschäden und bemerkenswerte Entladungsformen der atmosphärischen Elektrizität im Jahr 1901. Meteorologische Zeitschrift 20, 315.

A.Pühringer (1965) Comment, in: S.C.Coroniti (Ed.) Problems of Atmospheric and Space Electricity. Elsevier, Amsterdam, 460.

A.Pühringer (1967) Beobachtung eines Kugelblitzes im Salzkammergut und ein Erklärungsversuch. Wetter und Leben 19, 57.

S.Singer (1971) The Nature of Ball Lightning. Plenum Press, New York.

B.M.Smirnov (1987) The Properties and the Nature of Ball Lightning. Physics Reports 152, 177.

I.P.Stakhanov (1979) The Physical Nature of Ball Lightning. Atomizdat, Moscow (in Russian).

I.P.Stakhanov (1985) Abo he Physical Nature of Ball Lightning. Energoatomizdat, Moscow (in Russian).

E.Trapp (1948) Kugelblitzerscheinung in Wien. Wetter und Leben 1, 274.

M.D.Vernon (1971) The Psychology of Perception. Penguin Books, Harmondsworth, Middlesex, England.

W.von Haidinger (1868a) Elektrische Meteore am 20.Oktober 1868 in Wien beobachtet. Sitzungsberichte der mathematisch-naturwissenschaftlichen Klasse der k.u.k.Akademie der Wissenschaften, Wien, II, 58, 761.

W.von Haidinger (1868b) Die südwestlichen Blitzkugeln am 20.Oktober 1868. Sitzungsberichte der mathematisch-naturwissenschaftlichen Klasse der k.u.k.Akademie der Wissenschaften, Wien, II, 58, 955.

W.von Haidinger (1868c) Ein kugelförmiger Blitz, am 30.August 1865 gesehen zu Feistritz bei Peggau in Steiermark. Sitzungsberichte der mathematisch-naturwissenschaftlichen Klasse der k.u.k.Akademie der Wissenschaften, Wien, II, 58, 1045.

G.Zapletal (1970) Kugelblitz Schlins-Jupident. Wetter und Leben 22, 193.

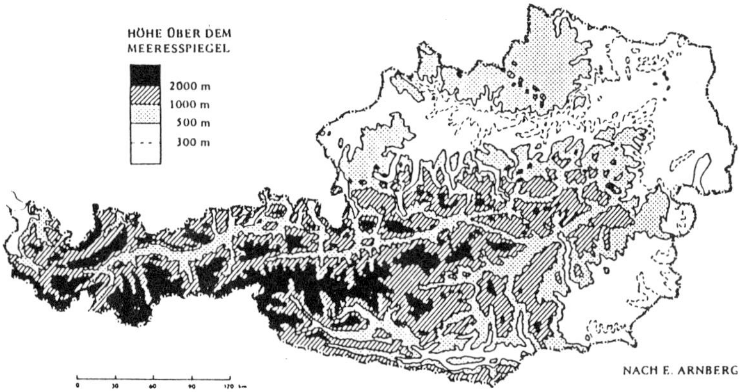

Figure 1a: Austria - heights above sea level (Austrian Central Statistical Office)

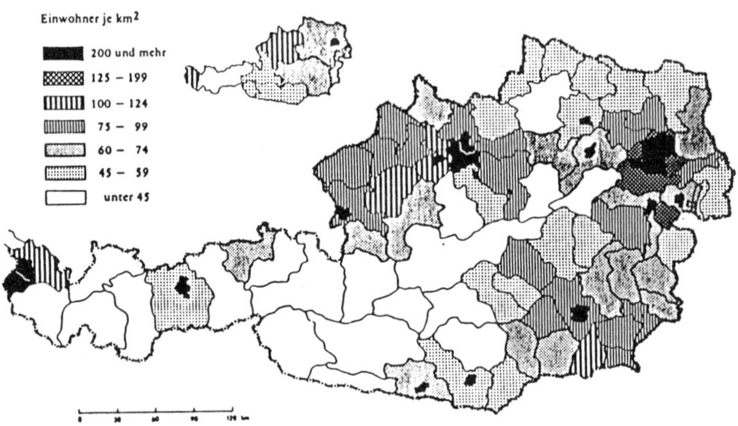

Figure 1b: Austria - inhabitants per square kilometer and county (Austrian Central Statistical Office)

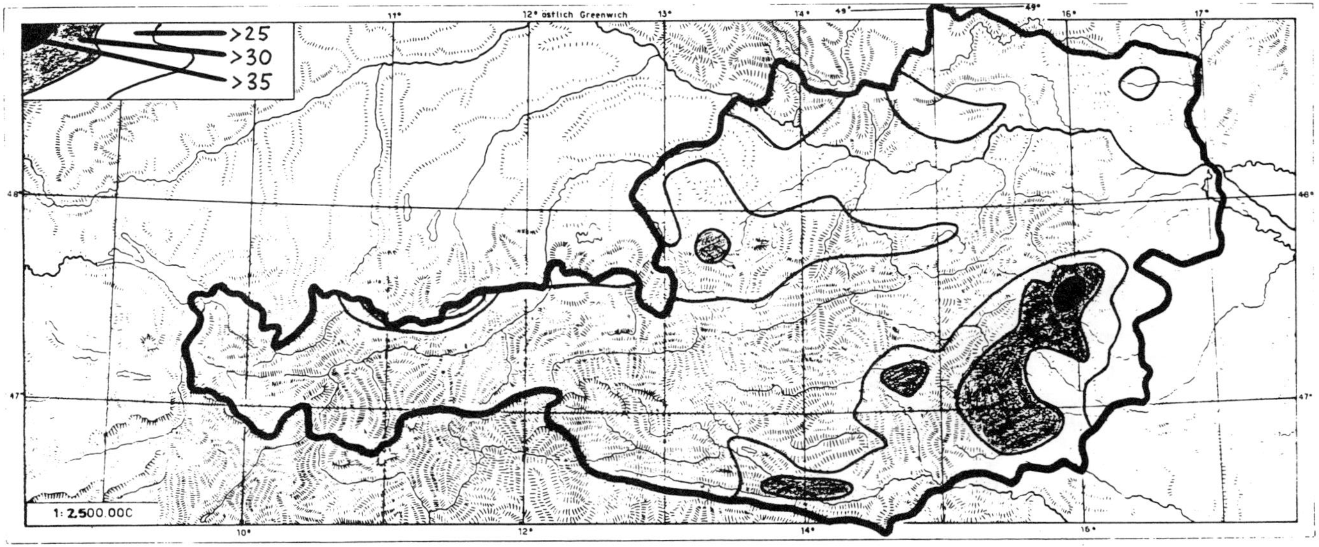

Figure 2: Austrian thunderstorm maxima (more than 25 days/year), 1901-1950 (Central Meteorological Office, Vienna)

Figure 3: Austrian BL folklore

Skopniak,
das feurige Schaubmandl

294. DIE LALLENDE KUGEL

Babbling ball (saga from Vienna area)

Es gilt als eine alte Erfahrung, daß gute Menschen hereinbrechendes Unglück ahnen und sich darauf vorbereiten können. Die bösen hingegen werden jählings vom Unheil überrascht. Hat nun der Himmel Seuchen, Hungersnot oder Krieg über die Menschen verhängt, so verkündet er dies durch die lallende Kugel. Im Marchfeld hat man sie in der Türken- und Schwedenzeit, dann in den schrecklichen Pestjahren, in der Huzulennot und zuletzt im unseligen neuner Jahr[1] gesehen. Tage und Wochen vor dem einbrechenden Unglück rollt eine feurige Kugel in der Größe eines Kinderkopfes durch Straßen und Gassen. Sie stößt Laute aus wie ein Menschenmund. Einige wollen ihr Lallen verstanden haben. Es ist eine Weissagung für die Frommen, wie lange der Jammer andauern und wie er enden wird.

[1] 1809, als die Franzosen in Österreich Krieg führten.

Skopniak, fire-dwarf (saga from Carinthia)

Figure 4: BL reports per pixel

and thunderstorm frequency in Austria

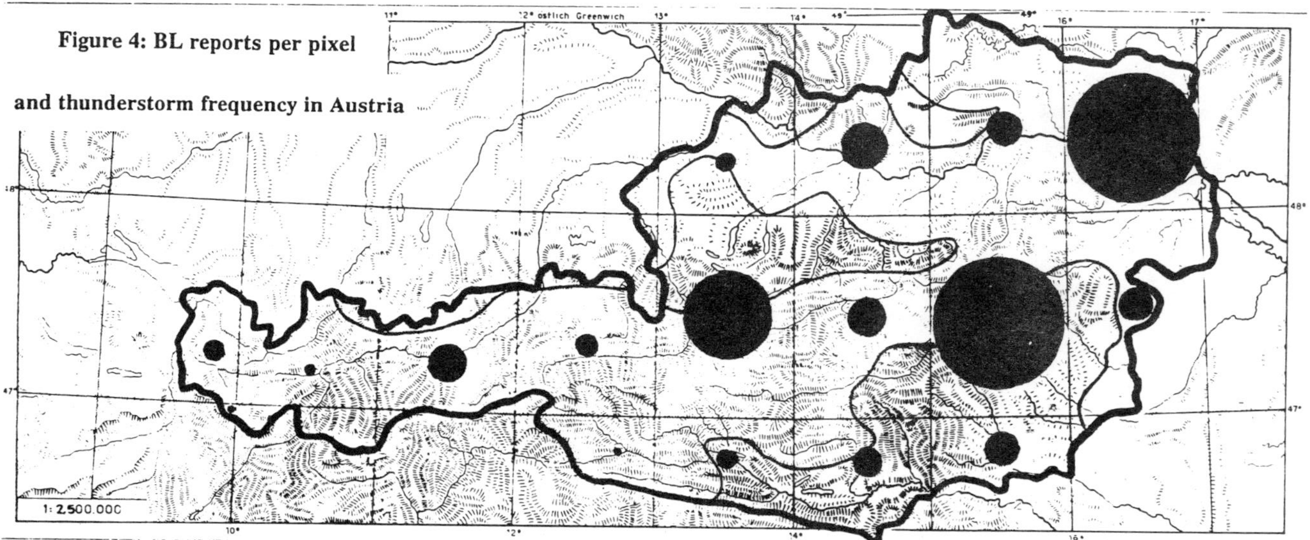

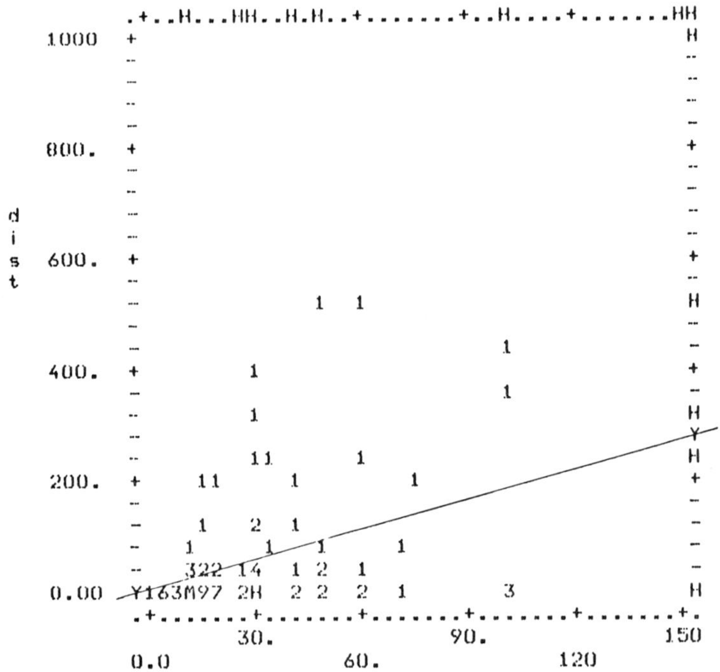

N = 114
R = .404
P < .001

--REGRESSION LINE--	-RES.MS-		MEAN	S.D.
Y=-3.3971 +1.8779*X	9205.3	X	27.640	22.474
		Y	48.509	104.43

groesse VERSUS dist (30 VS. 16)

Figure 5: Diameter (x-axis) versus distance (y-axis) linear regression

Table 1: Phenomenology of Austrian BL reports

Observers

- 52% female, 48% male (General population = GP 1981 53 : 47%)
- mean observing age 34 years
- mean reporting age 60 years
- mean "delta" 26 years
- 25% pupils, 26% employees, 10% academics (GP 1986 4% academics)
- 88% reported good eyesight (50% of cases; GP 1987 63%)
- emotional reactions: 60% calm/interested, 40% frightened

Situation, Environment

- 49% one witness, 50% two to ten, 1% more
- 1950, 1963, 1972, 1976, 1978 more than 3% of cases each
- June, July, August 79% of all reports
- 3-6 P.M. 25%, "afternoon" 18% of reports
- Vienna, Styria pixel each 20%, Salzburg 13% of cases
- 0-500 m sea level 57%, -1000 m 84%
- observer in settlement 85%, outside 15%
- observer indoors 64%, outdoors 36% (5% car, 1% plane)
- 50% less than 5 m distance (median 5,5 m), -100 m 76%, -1000 m 90%
- thunderstorm same time 72%, none 6%
- precipitation same time 53%, none 38%
- lightning stroke shortly before 39%, none 57%
- static electricity same time 10%, none 85%

Austrian "core phenomenon"

- single object 92% of cases
- duration: 1-5 seconds 72%, -10 sec 88%, -30 sec 97% (median 3 seconds)
- object in settlement 81%, outside 19%
- object outdoors 65%, indoors 35%
- object at ground level/near building 93%
- primary motion horizontal 50%, downwards 25%, complex, stationary each 12%
- secondary motion: hopping 6%, vibrating 4%
- contact to bad conductor 46%, to good conductor 37%, "free" 17%
- 91% round, 6% oval
- mean diameter 30 centimeters (median 25 cm)
  1-5 cm 6%, -10 cm 25%, -20 cm 48%, -30 cm 74%, -50 cm 87%, -100 cm 99%
- sharp outline 74%, fuzzy 19%, core + halo 7%
- red-orange-yellow 60%, blue-violet-bluewhite 16%, white 14%, variable 7%
- not blinding 82%, blinding 15%
- effects noticed: odor 14%, sound with object 15%, sparks 14%, tail 8%, rotation 22%, size variation 2%, "residue" 23% (mostly ordinary lightning effects)
- formation noticed: 25% of observers
- terminal explosion 39%, disappearing without sound 36%

Table 2: Significant Spearman rank correlation coefficients

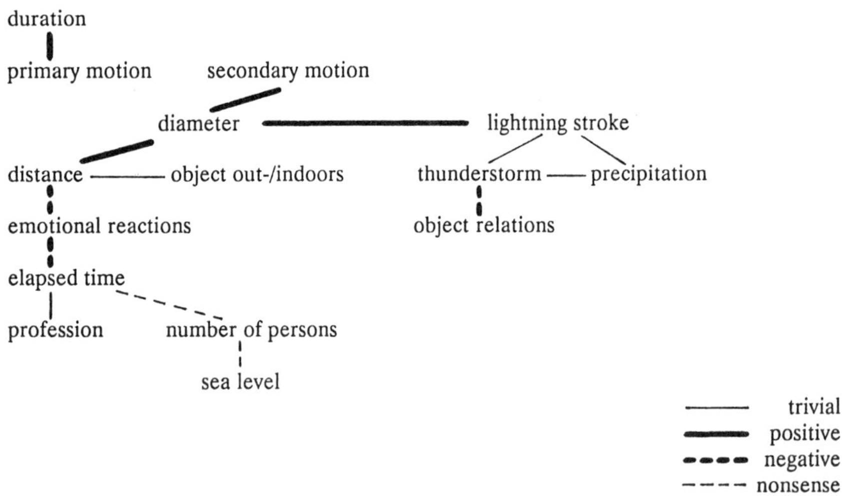

Table 3a: Austrian BL cases approximated by a Poisson distribution

150 cases; 16 pixels used x 78 years = 1248 pixel-years

Observations per pixel-year	observed	f(x) $\mu = 0.1$	approximated
0	1125	0.9048	1129
1	107	0.0905	113
2	11	0.0045	6
3	4 (A)	0.0002	0,2
4	1 (B)	0.0000	0

(A) = Vienna, Styria (B) = Styria

Figure 6: Possible relations observation - total duration

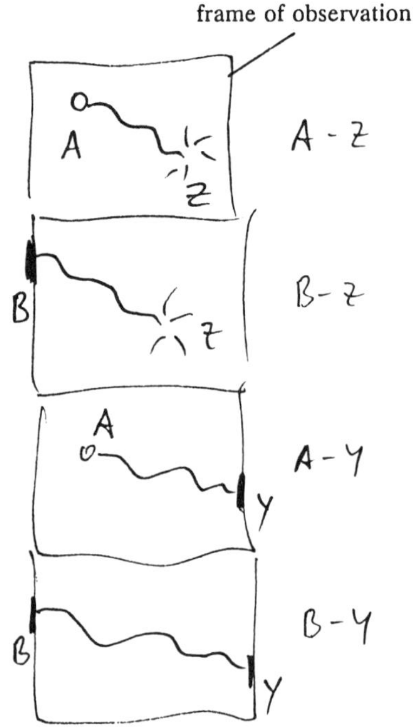

Table 3b: Austrian duration data observation - G=guessed, R=reconstructed (field)

		mean	st.dev.
A - Z	23 G	5.55	6.11
	3 R	**3.83**	**1.26**
B - Z	53 G	8.38	12.74
	2 R	**2.42**	**1.12**
A - Y	2 cases, not evaluated		
B - Y	16G	11.21	13.07

Table 4a: Multiple reporter evaluation

Locations different:		Locations identical:	
Observer A	4 d.v.	Observer E	4 d.v.
B	5	F	3
C	5	G	2
D	5	H	2,2

d.v. = number of different phenomena variables from case to case

In more equal 3 cases: duration, primary motion, object relation, diameter, outline, color, brightness, terminal explosion different variables.

Table 4b: Austrian BL and St.Elmo's Fire

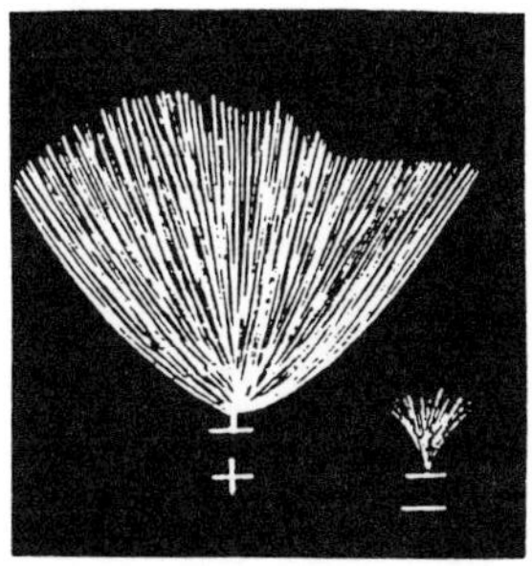

St.Elmo characteristics (Elster&Geitel, c.f.Gockel, 1925):

precipitation, object relation, color, simultaneous sound

Contact with good conductor - 8 cases
6 with precipitation; 4 white, 2 blue; 2 sound

From/to good conductor - 5 cases
3 with precipitation; 1 white, 1 orange, 1 violet; 1 sound

Near/on bad conductor - 16 cases
12 with precipitation; 2 white, 6 red, 1 orange, 2 yellow, 1 blue; 2 sound

Free-floating - 3 cases
3 with precipitation; 2 white, 1 yellow; 1 sound

Table 5: Comparison of some BL data bank evaluation results

Data bank	Dewan/McNally	Stakhanov	Lüdecke	Keul
Evaluator(s)	Dewan/McNally	St./e:Smirnov	Lüdecke	&Schwarzenb.
Publication year	1964,1966	1979,1985,1987	1982	1988
Data from	USA	USSR mostly?	Europe m.	Austria
Number of cases	513	over 1000	444 (*)	150
thunderstorm association	73%	62%	85%	72%
lightning stroke	85%		75%	39%
1-30 cm diameter	70% 2-41 cm	mean 22 cm	75%	74%
yellow-orange-red	47%	57%	60%	60%
less equal 10 sec	l.e.8 sec 78%	tau = 14 sec	80%	88%
maximum / year		83% JJA	58% summer	79% JJA
formation seen	32%		20%	25%
attached to/near/ origin at good conductor	63%		86% n=95	37%
explosive decay	sudden d. 73%	55%	63%	39%

Abbreviations: (*) = classical cases, including Brand and Galli data bank

tau = 50% of objects decayed

JJA = June, July, August

PHYSICAL PROBLEMS AND PHYSICAL PROPERTIES OF BALL LIGHTNING

George Egely
Central Research Institute for Physics
H-1525 Budapest 114, P. O. Box 49, Hungary

The physical properties of ball lightning /BL/ are not fully known and sometimes its important properties are doubted or ignored. The purpose of the paper is to review these properties and to pinpoint the significance of a given property for the various theoretical models.

Spatial and temporal stability

The usual shape of ball lightning is spherical /70%/, ellipsoidal/10%/, hollow spherical /10%/, pear-like and irregular /10%/. In the case of dim background illumination, a bluish or reddish aura covers the ball ($\sim$30% of the obsevations). No aura has been observed for hollow BL. Though in some cases the characteristic size of BL decreases as a function of time, it is usually stable even when the energy dissipation is significant, e.g. melting wet sandy soil into glassy slag (1); evaporating window pane (2,3).

The spatial stability is remarkable if BL passes through narrow orifices (like keyholes) since it recombines regaining its original shape (cases in (4, 5)).

Theoretical models based on surface tension cannot describe the stable pear shape nor the ellipsoid shape (6). The passage through orifices poses problems to ring current models (7), or rotating dipole models (8), because they are stable at a given radius. The samller radius of BL during its passage through orifices is an especially difficult problem and cannot be accounted for by these models. The remarkable spatial stability during interactions resulting in high energy dissipation poses a major problem to any internal energy source model like plasma models, ring current or rotating dipole models.

The stability of the colour of the phenomenon is observed

even in the case of interaction with different objects (made of copper, steel or aluminium); the colour is not altered by this interaction leading to high energy dissipation, like (1, 2, 3), do not lead to a change in colour. This casts doubt on models predicting the colour of BL on the basis of its temperature because the temperature should decrease in the case of internal energy models, and BL should be affected by interacting objects if it was base on plasma models. The colour of BL has been observed to change during the merging of smaller balls into a large one (5) so the change of colour is less likely to be due to chemical or plasma processes.

The temporal stability of BL is a well-known enigma. External energy source models would be able to explain the problem, but no such source has been found up to now. There are several internal energy source models that provide an explanation for the problem, but none of them has been verified experimentally. In fact, as Kapitza remarks (10), the fully ionized plasma sphere of a nuclear explosion recombines in a couple of seconds in the case of small surface/volume ratio. For BL, this ratio, is about four orders of magnitude higher, thus the cooling should be four orders of magnitude faster. The same argument is supported by the experimental data base of fusion research. The only exception is when there is chemical combustion: then the speed of the process is determined by the fuel and oxygen supply.

Energy Density, Total Energy, Damage Pattern

The energy density and total energy of BL vary over a wide range. Perhaps the highest energy density has been observed in a case described by Brand (11) where BL 3cm in diameter appeared in a kerosene lamp, then seriously damaged a church and several houses. The cases described in (1, 2) also have very high energy density ($\sim$600 J/cm^3) in Ref. (1), or 3.4×10^4 J/cm^3 in case 222 of Ref. 5 and there are at least five other similar obsevations in (5) and (11).

These values are too high for any plasma model or chemical

combustion model, as the maximum energy density for plasma models is about $\sim 2 \times 10^2$ J/cm^3 (12, 13).

These high energy denstity cases are substantiated by recent observations in (5) as well.

Apart from the high energy density, there is an additional important factor which could yield some insight into the physics of BL.

When the damage done by BL to different objects is analysed (14), it is apparent that known processes cannot have caused that particular type of damage to the given materials. The relatively uniform melting of a 35 cm thick sand layer (1) cannot be caused by heat radiation, high voltage current, ohmic heat or microwave radiation as was shown in the control experiment. The simulation of the evaporation of glass window panes was not successful (2), because mechanical stress caused by the uneven heating broke the glass window pane even at $\sim 20000^{\circ}$K.

When BL bores holes through adobe walls (5) the surface of the hole tends to be smooth, it has a uniform diameter without any trace of melting. The largest hole drilled by BL is about 60 cm deep and 20 cm diameter.

In one case Ref. (5) a 40 cm brick wall was penetrated by BL, but even though the penetration was roughly cylindrical.

Hot plasma balls or external E. M. radiation cannot cause such type of damage. The problem raised by these observations is that not only the BL itself but even <u>its damage patterns cannot be simulated under laboratory conditions</u>. This indicates also that a new physical phenomenon is to be sought as a principle behind BL.

The same contradiction appears when the electric charge content of BL is analysed, as well as the charge transfer. In principle, the electric charge of an average 20 cm diameter BL is in the order of $\sim 3 \times 10^{-6}$ Coulomb, otherwise the electric field on the surface exceeds 2.5×10^6 V/m, the breakdown voltage of air. Thus the electric potential of the surface is about 2.7×10^5 V. The electric energy stored in BL is $W \cong Q \times U$, that is, less than 0.8 J for our case. This amount of energy can melt ~ 0.3 mm^3 of aluminium or ~ 0.2 mm^3 copper. Even so there are several cases when all the electric wires in homes have been melted, sometimes this has meant as much as ~100 m of 1.5mm diameter aluminium wire. This damage is caused by electric current and not by heat conductions (because the process is almost instantaneous), nor by E. M. waves because bystanders do not feel heat and metal electric function box lids are not even warmed, only ejected from walls.

The major conclusion is that the electric charge and the energy requirement of the above process are unknown and this seems to contradict some of the conservation (charge, energy) principles. Yet, the above sequence of damage is quite usual: it can be found in every major collection of observations.

Usually it is difficult to estimate the total electric charge of a BL, but can be done sometimes indirectly with the help of the $Q = I \times t$ relation, or by the action of automatic fuses switched off by BL. Such a case was observed in Gyula, Hungary case 69 (5), when the current of an exploding BL switched off a 200 A automatic fuse for more than 1 sec. As the base current was about 120 A, the excess current had to be at least 80 A for more than 1 second, yielding 80 Coulomb total charge.

Another possibility for the estimation of charge and energy constant of BL is to apply the results of exploding wire research, so the charge and energy dissipation can be calculated (15).

Formation of Ball Lightning

There are two distinct means of formation, viz.

1. formation at the bifurcation point of streak lightning, or at the forking point of lightning.
   It is important to note that no BL formation has been observed without the possible change of direction of the streak lightning current, i.e. the bifurcation, when there are orthogonal electric currents, or in other words, strong inhomogeneous electromagnetic fields. These factors seem to be essential to the formation of BL.
2. spontaneous formation, when BL is formed in closed space /even in a Faraday cage/, under water, beneath the Earth's surface, in the soil, at the tip of metal conductors, or

in mid-air.

There is no difference in the behaviour of BL regarding the means and circumstances of the formation, they are identical in both cases. Spontaneous formation, however, poses the same problem regarding the conservation of mass and energy as has been discussed above.

Internal energy source models cannot explain this way of formation, and external energy source models cannot describe formation in a closed, shielded space. This is an important point showing again that there are fundamental problems with this phenomenon, because there are dozens of observations describing the spontaneous formation within a closed space.

The decay of BL also has two different forms:

1. Silent disappearance, usually a very fast process: the phenomenon disappears within a fraction of a second. Rarely, is the decay continuous.
2. Explosive decay when the explosion may release a significant amount of energy and electric charge, like in (1, 2) and several cases in (5) and (11).

In most cases the decay is explosive if there is interaction with a neighbouring object, though not always.

The colour, size, brightness, way of formation, etc. has no definite effect of the type of decay; none of the theories can predict it.

The movement of BL has several possibly important features:

1. In some of the observations rotation around a horizontal axis is described. The velocity of the movement is usually perpendicular to the axis of rotation. The angular velocity is steady, so it is not a gradually decelerating vortex-like behaviour. This persistent feature casts doubts on all plasma models, where friction with the ambient air should gradually decelerate the BL.
2. In a closed space (rooms) it moves around the walls along a horizontal plane and never along a vertical one. This feature casts further doubts on hot plasma models, because in that case the velocity of BL should have a vertical component as well. But this feature - the existence of BL in a closed

space - casts doubts on all external energy models as well, because E. M. radiation has been ruled out previously and other forms of energy cannot be transported through the wall without any trace.

3. There are some observations where BL passed through a thin glass window pane without harming it (5, 11, 16). (Usually it heats up the glass so it is broken by thermal stress). This phenomenon casts doubts on most plasma models as the regeneration of BL on the other side of the glass cannot easily be explained, though Powel and Finkelstein (3) incorporated this possibility into their model.
4. The movement of BL is apparently guided by electric fields as BL is attracted by electrically conducting objects and permanently attracted by grounded conductors.
5. There is no evidence that wind has any effect upon the movement of BL, though sometimes their velocities could be parallel, but this is only by chance. BL having an internal energy source should be influence by the wind, consequently there is a further reason to doubt the internal energy source as a possible solution to the problem.

There above list of contradictions partly explain the reason for the lack of progress, for the lack of a solution. This, however, does not rule out a solution: it only shows that a novel type of solution is to be sought for the enigma of ball lightning.

References

1) Dimitriev M. T., Bahtin B. I., Martinov V. I. Zh. Techn. Fiziki (USSR) Vol. 51. pp. 2567-72 (1981)
Eng. translation: Sov. Phys. Techn. Fiz. (USA) Vol. 26, pp. 1518-20 (1981)
2) Kolosovskii, O. A.: Zh. Techn. Fiz. (USSR), Vol. 51, pp. 856-858 (1981)
Eng. transtlation: Sov. Phys. Techn. Phys. (USA) Vol. 26, pp. 510-511 (1981)
3) Powel, J. R., Finkelstein, D.: American Scientist, Vol. 58, pp. 262-279 (1970)
4) Stahanov, I. P.: O fizitseskoj Prirode Sarovoj Molnii Moscow, Energoatomizdat, 1985.
5) Egely G.: Hungarian Ball Lightning Observations. Research

Report, Central Res. Inst. for Physics, Budapest, Hungary, KFKI-1987-10/D.
6) Brovetto, P., Maxia,VV.: Journal of Atmospheric and Terrestial Physics. Vol. 38, pp. 921-934 (1976)
7) Moigno, F., Cosmos Vol. 14. p. 672 (1859)
8) Endean, V. G., Nature, Vol. 263, 753 (1976)
9) Barry, J. D.: Ball Lightning and Bead Lightning, Plenum Press, New York, 1980.
10)Kapitza, P. L.: Dokladi Akademia Nauk SSSR, Vol. 101, pp. 245-248 (1955)
11)Brand, W.: Der Kugelblitz Henri Grand, Hamburg 1923.
12)Barry. J. D.: Journ. Geophys. Res. Vol. 85, NOC7, pp. 4111-4114 (1980)
13)Kozlov, B. N.: Dokl. Akademii Nauk SSSR Vol. 238, pp. 61-62 (1978)
14)Egely, G.: Energy Transfer Problems of Ball Lightning. Research Report. Central Res. Inst. for Physics Budapest, Hungary KFKI-1986-13/D
15)Chace, W. G., Moore, H. K.: Exploding wires, Plenum Press (1959)
16)Uman, M. A.: Journal Atmospheric and Terrestial Physics Vol. 30., pp. 1245-1246 (1968)

STATISTICAL ANALYSIS OF THE BALL LIGHTNING PROPERTIES

Grigorjev A.I., Grigorjeva I.D., Shirjajeva S.O.
University of Yaroslavl', YAROSLAVL', USSR

As has repeatedly been stated in numerous scientific publications on BL the main source of the new information concerning this atmospheric phenomenon still remains the questioning of eyewitnesses (see for example [1-5] and the literature given there). The publications devoted to the theoretical and experimental BL modelling have not yet left the frames of hypothesing assumptions though one can notice a certain progress in this field.

The questioning of the BL eyewitnesses made by McNally [5], Imjanitov I.M. [2], continued by Stakhanov I.P. [3] and later on by the authors of the given work [7] made it possible to specify and widen the list of properties of this type of lightning as compared to classical notions [1,8].

Speculative conclusions caused by the insignificant number of BL observations were followed by an accurate statistical analysis on the basis of thousands of observation cases. This type of analysis was successfully applied for the 1st time in [3]. In the given work such BL parameters as lifetime, characteristic linear dimensions and velocity are being analysed according to the methods used in [3] on the base of 2082 observation cases. The data are also given concerning the frequencies of other BL properties such as: colour, brightness, shape, electrical and magnetic properties etc. within our collected observation material.

1. BL Lifetime.

Now there is no doubt about the fact that by its physical nature BL is a certain case of plasma. This however makes it more confusing to understand its relatively continuous autonomous existence which pretends to be a stumbling block for new models [3 - 6] . In this respect the question of BL lifetime estimation on the basis of the data processing of observation material acquires special importance.

As has already been stated in [3] the number of BL observations from the moment of its appearance up to its decay is insignificant. For this reason in [3] the BL lifetime was identified with the period of observation for extension of statistics.

In the present investigation such an identification takes no place as the number of observation cases has increased more than twice as compared to [3] . The analysis below is being made for three groups of BL descriptions in which: 1) BL was watched from the moment of its formation up to its decomposition (256 cases); 2) BL appeared promptly after the discharge of lightning nearby (hence one can assume BL to be formed as a result of the lightning discharge and, consequently, the former was watched nearly from the moment of its formation) and up to its decomposition (181 cases); 3) BL left the eyewitness's sight or the latter failed to fix the moment of its formation and BL appearance is not related to the lightning discharge (1522 cases).

For the further analysis the duration of observation periods can be divided into the following intervals (in seconds): 0 - 3; 3 - 5; 5 - 10; 10 - 15; 15 - 20; 20 - 25; 25 - 35; 35 - 50; 50 - 70; 70 - 100; 100 - 200. The calculation of the BL cases where BL observation was terminated after certain periods given above per 1000 events allows us to draw the curvs of probability of BL decomposition. The points corresponding to the centres of the intervals for 3 groups of BL cases are presented in fig. 1 by triangles, squares and circles correspondingly. The searched dependences can be approximated by the expression:

$$P = A t^{n} \exp(-K/t),$$

where the constants A, n, K can be easily found by the method of the least squares. For the first two groups of descriptions the following values are obtained: A = 0,59; n = 0,1; K = 0,8 sec; for the third one: A = 0,59; n = 0,1; K = 0,5 sec. Lines 1 and 2 in fig. 1 present the data of Stakhanov's (1) and McNally's (2) for which the following values are found: A = 0,31; n = 0,24; K = 1,29 s. and A = 1,007; n = 0; K = 1,41 s. respectively. While calculating the "experimental" points were accepted related to the centres of the corresponding intervals.

The fact that the dependences corresponding to the first 2 groups according to our data coincide testifies in favour of the assumption that the observation time in the second group can be identified with BL lifetime. At the same time the difference between the first two dependences and the third group reveals the fact that it is not quite accurate to assume the observation time for BL lifetime in all the cases as it was assumed in [3].

Following the methods used in [3] we draw the diagram of dependence on time of the rest number of BL cases assembled by 1000 events (attributing the "experimental" point to the right ends of the intervals) in a semi-logarithmic scale. Hence it is possible to find out that the law of decomposition is well approximated by three exponents whose periods of decay differ greately:

$$N = N_0 \{1 - \alpha_1 \exp(-t/\tau_1) - \alpha_2 \exp(-t/\tau_2) - \\ - \alpha_3 \exp(-t/\tau_3)\},$$

where N is the number of BLs terminating their existence during the period of time t out of the total number N_0 (N_0 = 1000), α_1, α_2 and α_3 are the parts of BLs decomposing during the particular periods τ_1, τ_2 and τ_3 correspondingly ($\alpha_1 + \alpha_2 + \alpha_3 = 1$). For the first gruop we obtain: α_1 = 0,59, τ_1 = 3 sec; α_2 = 0,26, τ_2 = 29 sec; α_3 = 0,15, τ_3 = 213 sec. For the second group; α_1 = 0,58, τ_1 = 2,9 sec; α_2 = 0,28, τ_2 = 32 sec; α_3 = 0,14, τ_3 = 217 sec. For the third group: α_1 = 0,6, τ_1 = 2,5 sec; α_2 = 0,27, τ_2 = 26sec; α_3 = 0,13, τ_3 = 400 sec. The division of all the BL cases into three types with various observation time can be done quite distinctly as it is seen from fig. 2. The fact that the characteristic periods of decomposition as well as BL general shares of the first two types are approximately the same in all the three groups deserves attention. It is of interest that τ_1 decreases from the first group to the third one (though this change is within the error $\varepsilon_t \approx 0,57$, see the appendix).

The law of BL decomposition in a collection of 1000 events throughout the observation material obtained by Stakhanov I.P. is characterized by the existence of two BL groups: α_1 = 0,57, τ_1 = 11 sec; α_2 = 0,43, τ_2 = 54 sec [3]. We have repeated Stakhanov's calculations in order to clear up the possibility to pick out in his collection of data 3 groups of BLs, similar to those obtained in the present analysis. We have to admit however that the division into two groups suggested by Stakhanov corresponds to the real facts used in [3] in an optimum way. The results of the similar analysis for the data after McNally and Raile are also presented in [3] . In McNally's collection of data two groups of BLs can also be selected: α_1 = 0,81, τ_1 = 3,5 sec;

α_2 = 0,14, τ_2 = 64 sec, while Raile's material allows to pick out only one BL type: α = 1, τ = 14 s. Having compared the results obtained during the analysis one can see that the group of shortliving BLs after McNally's data reveals the characteristics (α_1 and τ_1) which appear to be the nearest to our data. The absence of such a group (with α_1 = 3 s.) in Stakhanov's selecting analysis seems to be the result of his having used the method of questioning in the process of collecting the information and of too coarse division into intervals with small periodas of observation: the first three intervals of observation time in [3] were: $\leq$ I s; 1 - 5 s; 5 - 10 s. Under such division it was just impossible to pick out a BL group with the characteristic period of decomposition ≈ 3 s.

As has been stated above, while analysing the laws of decomposition of a collection of 1000 BL events the "experimental" points were attributed to the right ends of the corresponding intervals of time. This is in accordance with the essence of the statistical analysis of BL events left in the collection by a certain moment. In this connection the problem ofsensitivityof the data obtained has arisen, i.e. the sensitivity of α_i and τ_i to the position of "experimental" points within the interval. As the analysis has shown, the meanings of α_i - the percentage of BLs of various types in the collection practically does not depend on the positions of "experimental" points within the interval (i.e. on which end of the interval they can be attributed to), while the characteristic periods of decomposition τ_i for all the BL types change proportionally to the displacement of the points within the interval. In other words, the relative error in defining the time of BL existence (observation time), which has the valeu $\varepsilon_t \approx$ 0,57 according to the appendix, will lead to the same relative error in defining characteristic periods of decomposition τ_i. Summing up all these facts the data obtained above can be interpreted in favour of the existence of three BL types: shortliving with $\tau_1 \approx 3$ s (after McNally's and our data), BLs of the mean period of existence with $\tau_2 \approx 30$ s (after Raile's and our data) and longliving BLs with τ_3 = 100 - 200 s (after the data of Stakhanov's, McNally's and ours).

2. BL Shape, Structure and Dimensions

The question of BL shape is not an idle one. The reason for this lies in the fact that the BL capacity to change the shape during its existence or in interaction with an obstacle will facilitate to clear up its structural mechanism. The values of probability of various BL shapes according to our data as well as to Stakhanov's and Rayle's are presented in table 1.

There are 134 reports of BL changing shape in the process of its existence. In 18,66% of the cases described a ball turned into a band, in 11,19% - a band became a ball, in 2,98% BL flattened while jumping, in 16,42% of descriptions the BL surface flactuation was reported, in 8,96% of cases it changed the shape immediately before the discharge on a conductor: it stretched towards the conductor, in 41,79% BL form changing was reported without any details.

The question of BL form and dimension adjoin that of BL having a semi-transparent shell mentioned in 226 reports and which, if available, can disfort the estimation of BL visual dimensions. The average thickness of the shell in all the cases described is ≈ 3 cm, the colour of the shell being distributed with probabilities given in table 2 .

The reports of BL visual dimensions may be influenced by the observation of a BL tail reported in 119 cases.

In 143 cases ($\approx 6,9\%$) BL was observed having a certain inner structure: bright luminous dots chaotically moving within the BL volume or clusters of threads brighter against the main background.

For the purposes of statistical analysis of the BL dimension distributions all the reports of BL diameters are grouped on diapasons (in cm.): 1 - 3; 3 - 5; 5 - 10; 10 - 15; 15 - 20; 20 - 25; 25 - 30; 30 - 35; 35 - 40; 40 - 45; 45 - 50; 50 - 70; 70 - 100, > 100. Fig. 3 gives the statistical function of the diameter distribution based on 1796 events described. The analytical dependence optimally approximating the statistical data by the method of the least squares looks like:

$$P = 1 - 1{,}2\, d^{-0{,}1} \exp\left(-\frac{d}{d_o}\right),$$

where d_o = 20 cm.

In fig. 3 curve 1 is the density of BL distribution normalized on a unit for diameter, curvs 2 - 4 are the same density of distribution according to Stakhanov's data (1005 events), McNally's (446 events) and Rayle's (98 events) [3] respectively. The qualitative coordination of these dependences is evident, though in our collection of data the position of maximum is displaced towards smaller significance and falls on 7,5 cm. Curvs 2 - 4 in [3] were approximated by the expression:

$$f(X) = \frac{X}{X_0^2} \exp\left(-\frac{X}{X_0}\right)$$

Unfotunatly according to our statistics the density of distribution of BL diameter cannot be described so easily and has a more complicated analytical description:

$$\mathcal{K}(d) = \mathcal{K}_* + d\left(\varkappa_1 e^{-\frac{d}{d_0}} - \varkappa_2\right); \qquad \mathcal{K}_* = 10^{-3}, \quad d \leq 200\ cm$$

($\varkappa_1 = 0{,}013\ cm^{-1}$; $\varkappa_2 = 6 \cdot 10^{-6} cm$; $d_0 = 1/0{,}12\ cm$). This is obtained by the treatment of statistical points according to the method of the least squares.

Table 3 presents the values of the most probable diameter d_{max} (for which the density of probability is maximum), as well as the average value of BL diameter defined as

$$\langle d \rangle = \int_0^\infty d\, y(d)\, d\, d$$

and obtained by numerical integration.

In 71 cases ($\approx$ 3,5%) eyewitnesses point out the BL ability to be divided into smaller units quite spontanuously without any interference of an outer force, the number of secondary BLs being less than 6 in 45,1% and more than 9 in 54,9%.

3. Moving Character

BL moving features are illustrated by table 4. The wind influence on BL moving character was shown by our data in 52 cases, BL being carried by the wind in 42 cases, BL being indifferent to the wind in 10 cases including 4 cases of BL moving against the wind.

The BL velocity distribution based on 1286 observations is given in hystogram fig. 4 as a solid line. The same hystogram gives Stakhanov's data [3] as a dash-dotted line.

Besides BL changed its velocity while moving in 112 cases according to our statistics.

In 70 cases BL jumped after the struck against the earch like a rubber ball. This seems to show the fact of BL having surface tension. On the other hand in 18 observations it used to jump back from a vertical wall.

In 104 descriptions the BL ability to penetrate through a hole smaller than BL diameter is reported, BL changing its form in 30 cases and remaining unchanged in 12 cases.

4. Colour, Brightness, Thermal Radiation

The problem of BL thermal properties is closely connected with that of the energy accumulated in it, of the way of energy storage and possible ways of energy release. From 383 observations received by the authors when the eyewitnesses payed attention at the heat radiation in 319 ($\approx$ 83%) cases BL actually did not release any heat and in 64 ($\approx$17%) there was certain warmth sensation. The data differ greately from those of Stakhanov's [3] who stated the warmth sensation only in 4,9% of 1293 events. Moreover when the distance between an eyewitness and BL decreases the number of warmth sensation cases increases. Unfortunately our statistics do not confirm such a correlation: 232 persons reported of BL warmth properties at a distance of a meter or less than a meter and only 14,2% sensed the warmth. However the whole collection of the heat releasing BL cases being rather small the statistical investigation gave no distinct results. The colour distribution of the heat releasing BLs is approximately similar to that of the general number of BL observations but the percentage of the "white" colour among the heat releasing BLs was on a third less. The brightness analysis showed that the heat releasing BLs whose brightness coinsided with that of an electric lamp of power 100w take 80% while the BLs of such brightness take only 69,4% in the total collection of events. The diameter distribution of heat releasing BLs is the same throughout the took collection of cases. The availability of inner structure was marked in 10 out of 64 cases. This corresponds to the probability twice greater than for the whole collection of events.

In three cases BL caused the burns similar to those of intensive ultra-violet radiation.

Table 5 shows the frequencies of variously coloured BLs accurance. When eyewitnesses observed intermediate colours: red-orange, orange-yellow the corresponding description refered to the long-waved component of colour. The same table contains Stakhanov's data [3] as well as the average probability calculated by Smirnov B.M. [6] according to the collection of data after Stakhanov, Rayle, Charman and McNally.

For the estimation of BL brightness the observes were offered the following standards: a candle, a (kerosene) oil lamp, the moon, an electric lamp of power 15w, 25w, 100w, 150, 200, 300. Besides there were such brightness characteristics in the descriptions as: "dull", "grimmy", "bright", "dazzling bright", "transparant", "semi-transparant". To compare the reports made by our eyewitnesses with Stakhanov's data who used as a standard only electric lamps, we divide our cases into 6 groups approximately corresponding to the brightness ranges suggested in [3] . Table 6 illustrates the relations between the standards of brightness in [3] and those used in the given work. The probabilities of a given range of brightness are shown in colomns 3 and 4 of table 6. It must be noted that the suggested division is rather conventional and an error in the estimation of brightness is assumed. But the standards used in [3] are no less conventional. Indeed, electric lamps of power 10w, 20w, 300w, 500w seem to be too exotic to be recommended as standards. Besides the sensation of brightness of a particular lamp depends on the availability of other sources of light (e.g. BL observation differ in different hours in the daytime and at night). However the standard lamps system gives an opportunity for further numerical statistical analysis of the errors assumed and consequently we are going to use it.

5. Conditions of BL Appearance and Observation.

In 35,35% out of 1932 descriptions where the place of BL appearance was fixed it was observed in town, in 64,19% - in the countryside, in 0,31% (6 descriptions) - in the clouds and in 0,16% (3 descriptions) - inside airplanes. More detailed indication of BL observation places based on 1984 descriptions is contained in table 7.

The correlation of BL appearance conditions with thunder is illustrated by the data of table 8.

BL was born in the presence of an observer in 286 cases according to our collection of data (see table 9).

The distribution of minimum distances between an observer and BL is presented in table 10.

Table 11 illustrates the peculiarities of BL decomposition in 1907 cases.

6. Correlation between the Intensity of Thunder Activity and the Frequency of BL Observation.

The problem of meteorological conditions of BL appearance is of interest for clearing up the formation mechanism and the study of BL structure. The strong dependence on thunder conditions stated, it would give an opportunity to considerably restrict the range of possible BL structural models the total number of which now exceeds two hundred [3, 4]. This task is particularly important if we take into account the fact of BL appearance in fine weather: ≈9,77% according to our data, 13% of 1006 descriptions according to [3]. The problem of treatment of these facts still remains open: they may testify either to the possibility of non-thunder BL origin or to slight reliability of eyewitnesses' reports of the meteorological situation in the moment of BL observation. However let us turn to the results of the statistical analysis.

1. Consider the standard linear correlational analysis of the distribution data on the frequency of BL observations per month (1713 descriptions all in all), see table 12 as well as the data of thunder recurrence given in [9]. We will find the value of correlational coefficient between these two events.

$$\eta = \frac{\sum_{i=1}^{n} \Delta X_{1i} \Delta Y_{1i}}{\sqrt{\sum_{i=1}^{n} \Delta X_{1i} \sum_{j=1}^{n} \Delta Y_{1j}}}$$

which equals approximately 0,79 correct with confidential probability 0,98. The straight line of the regression X_1 - BL observation frequency (in percentage of the total number of cases) on Y_1 - thunderstorm recurrence (in percentage of the average annual number) looks like

$$X_1 = 8,3 + 0,2\,(Y_1 - 8,3). \qquad (1)$$

The coefficient of the same correlation calculated in [3] is approximately equal to $\approx 0,83$ and the corresponding regression line looks like:

$$X_1 = 8,4 + 0,19(Y_1 - 8,3). \qquad (2)$$

Here we have quite a satisfactory agreement of our calculation and the results of data processing in [3]. The following fact acquires attention: the average annual thunderstorm number per month - 8,3% (out of the total number) completely coincides with the average annual BL observation number per month: here - 8,3%, in [3] - 8,4% (out of the total number as well).

2. Let us study the obtained data of the distribution of BL observations during the daytime and at night (see table 13) as well as the statistics of thunderstorm number recurrence during the day. We shall find the correlation dependence between these two events with the value of correlation coefficient equal to $\approx 0,88$ correct with confidential probability 0,95. The regression line X_2 - of the number of BL observations on Y_2 - the thunderstorm recurrence in various periods of the day (in percentage out of the total number of cases) is expressed by the equation:

$$X_2 = 12,5 + 2,2(Y_2 - 12,5).$$

Here again one can notice the coincidence of the average BL observation number for all the intervals - 12,5% with the average thunderstorm number being 12,5% as well.

3. Let us consider the statistics of the thunderstorm duration (in hours) in various latitude-longitude ranges throughout the territory of the USSR [9] and our data of BL observation number in the same latitude-longitude intervals (see table 14). We shall find the value of the correlation coefficient between these two events which is equal to $\approx 0,65$ correct with confidential probability 0,99. The regression line of BL observation number X_3 on the average thunderstorm duration in various latitude-longitude intervals looks like:

$$X_3 = 96 + [4,2(Y_2 - 25)]$$

where the square brackets mean that the integer part of the number within them is taken.

The correlation coefficient obtained seems remarkably large if we take into account the strong hampering influence of the uneven distribution of eyewitnesses throughout the territory of the USSR as a result of the uneven distribution of the population (according to the statistical data about 3/4 of all BL cases were observed in populated areas: 50,15% indoors and 24,56% outdoors).

4. On studiyng 1212 BL cases on the territory of the USSR since 1920 till 1965 the value of the coefficient between BL observation number during a year and the solar activity characterized by the average annual value of Volf's number was found which is equal to 0,56 correct with confidential probability 0,99. This problem was put in [10] and solved in [11] .

The idea of the influence of the solar activity on the weather-climatic conditions of the Earth arouses no objections nowadays and is confirmed by a great number of scientific publications on this problem (see e.g. [12] and the literature given there). But this influence is exercised indirectly through a number of intermediate mechanisms and it is rather difficult to reveal it in each particular case.

The first attempt to reveal the correlation dependence between BL observation number in the territory of Holland and the solar activity for the period since 1920 till 1965 was made in the present work [10] .However in [10] the searched correlation dependence was spoken of on the basis of the external likeness of the curvs of time dependence of the observation number and Wolf's numbers while the statistical data processing was not presented. Arabagy [10] was searching for correlation dependence between BL observation fequency and the solar activity to confirm his model which he had proposed earlier and which connected BL origin with the corpuscular solar radiation. Still there were not enough data used by him to state the fact of existence of the searched correlation dependence. The value of the correlation coefficient between the average annual value of Wolf's number and BL observation frequency calculated by the authors of the present work according to the data in [10] appeared to be equal only to 0,06.

The existence of searched correlation was found out on the basis of the reports of BL appearance on the territory of the USSR collected by the authors of the given work. In fig. 5 curve 1 presents the annúal distribution of BL observation frequency throughout the territory of the USSR. The distribution according to Arabagy's data for Holland is represented by curve 2; according to the general (summerized) statistics it is curve 3. The annual distribution of Wolf's numbers from [13] is represented by curve 4.

The correlation analysis has shown that during the period since 1920 till 1965 the correlation coefficient between BL observation frequency and the solar activity throughout the territory of the USSR equals 0,56. And the regression line of BL observation frequency X_4 on Wolf's numbers looks like:

$$X_4 = 26 + [0,1\,(Y_4 - 62)].$$

It is quite natural to treat the statistically valuable result thus obtained as the influence of the solar activity on the frequency of BL appearance indirectly, through the intensity of thunderstorm activity.

The results presented above in the authors' view convincingly demonstrate the attachment of BL observation to thunder conditions. As for some eyewitnesses' reports of BL appearance in fine weather they do not seem to correspond to real facts. As we do not suspect on the part of our reporters an attempt to mislead us, one could try to explain the mistake by the following: 1) the observers might interprete as "fine" any kind of weather when the sun shines, 2) the influence of thunder is felt at a distance of approximately 10 km thanks, for example, to the opportunity of overvoltage waves to spread along electrical transmission lines (it must be mentioned here that 68,53% of 286 cases of BL formation in the presence of eyewitnesses took place on a conductor, the data being in conformity with those of Imjanitov's and Stakhanov's [3]); 3) As a rule the observers' information comes with delay, some years after the event, and the meteorological conditions are not always reconstructed precisely.

7. Correlation Dependences between Characteristic BL Linear Dimensions, Lifetime, Brightness and Colour.

In conformity with the facts presented above rather a complete information has been collected to the present day concerning the properties of a mean BL [1-7] . The first attempts have also been made to investigate the relationship between BL observation frequencies and thunder conditions and to find out the correlation ties between separate BL properties (see e.g. [3,11,14-16]). The latter problem is the subject of the part below.

1. While searching for correlation between separate BL properties first of all one has to face with a low degree of reliability of the data of BL properties reported by eyewitnesses. This is caused by the peculiarities of people in estimation of spatio-temporal intervals, colours and various kinds of brightness of objects. The fact that they have to do this estimation after rather a long period of time is also the reason for possible inaccuracy These sources of errors in the collected information as well as some others have been analyzed and are contained in the appendix. There the tables of relative errors in the values of BL characteristics accepted by the authors are presented which were discovered by the statistical data processing. However owing to the shortage of the statistical material we shall use further the average values of relative errors in all possible ranges of BL characteristics. Thus for BL linear dimension the relative error in the reported data is equal to: $\varepsilon_d = 0,14$; for observation period it is $\varepsilon_t = 0,3$; for brightness $\varepsilon_b = 0,3$ (if the standard of electric lamps accepted by Stakhanov [3] is used); for colour $\varepsilon_c = 0,025$ (if the colour is characterized by the corresponding electromagnetic wave-length).

The searching methods of correlated properties X and Y from the ranges i and j were correspondingly described earlier in [14] and come to the following. For the properties X_i and Y_j probabilities P_i and P_j in the total collection of BL descriptious are to be calculated. Then the probability P_{ij} of BL possessing the properties X_i and Y_j is to be searched throughout the total collection of data. On condition $P_{ij} > P_i P_j$ it is accepted that this combination of properties is not accidental and hence there is correlation between X_i and Y_j .

Particularly, to establish the dependences for each characteristic of brightness and colour stated above, the characteristic BL linear dimension and lifetime correlated with these properties were selected and the average values of dimension and lifetime were calculated. The results of this calculation are presented in fig. 6 - fig. 9.

2. To analyse the correlations, all BL descriptions were divided into 6 groups (see the appendix) according to the degree of increas in brightness on the basis of comparison with the brightness of electric lamps of different power [3] : 1) weaker than a lamp of 10 w, like a lamp of: 2) 10 - 20 w; 3) 20 - 50 w; 4) 50 - 100 w, 5) 100 - 200 w; 6) brighter than a lamp of 200 w. For each group the average values of BL characteristic linear dimension and lifetime were found according to the methods mentioned above. The points thus obtained are given in fig. 6 (the correlation dependence of characteristic BL linear dimension and brightness) and in fig. 7 (the correlation interdependence of lifetime and brightness) where a degree of brightness through the equivalent brightness of the established six groups is laid off on the axis of abscissa. The probability of a degree of brightness of the given group occuring throughout the total collection of BL descriptions is marked in round brackets under each graduation of the abscissa axis: according to the author's data it is the upper digit, according to the data [3] it is the lower one. The quantity of correlated events corresponding to each point in the diagram is given near the point in round brackets.

The correlation data of BL linear dimension and lifetime with its brightness given in fig. 6 - fig. 7 are found from the descriptions of BL observation at a distance not more than 20 m. The reports of the eyewitnesses who watched BL at a distance more than 20 m are thrown aside as it is rather difficult to estimate the brightness of a far-distanced objects and so the eywitnesses' reports are considered of low reliability.

It is obvious for physical reasons that the dependence of BL radiation intensity on its diameter are to be described by the expression: $\mathcal{J} = \mathcal{J}_0 d^n$. If radiation comes from BL surface then $n = 2$, if the whole volume of BL radiates , then $n = 3$. In intermediate cases $2 < n < 3$. In this connection the "experimantal" points in fig. 6 were extrapolated by the stated dependence for which after processing by the method of the least squares we got: $\mathcal{J}_0 = 5 \cdot 10^{-2}$ w/cm , $n = 2{,}4$. The changes in the positions of the points in fig. 6 within the confidential intervals (see the appendix) causes the changes of rated values of the parameters of the searched dependence in the intervals: $4\cdot10^{-2} \leq I_0 \leq 8\cdot10^{-2}$ w/cm^2; $2{,}26 \leq n \leq 2{,}45$. This indicates the fact that BL radiation comes out of some surface layer with thickness $h \approx 0{,}2d$. This fact is in conformity with the data stated above of BL having a semi-transparent cover. From the physical point of view the obtained result means the following: BL seems to be a kind of plasma with optical thickness of the order of decimal parts of visual dimension.

3. The possible BL colour characteristics were accepted as the following: violet dark blue, blue,green, yellow, orange, red, white. For such often reported colours as: red-orange, yellow-orange, the colour corresponding to the longer electromagnetic wave was chosen during the data processing. In fig. 8 - fig. 9 the correlation dependences of BL characteristic linear dimension and lifetime with its colour are presented. The colour characteristics are laid off on the abscissa axis in the order of wave length increase. The probability of each colour characteristic in the total collection of descriptions is given in brackets under the characteristic: according to the authors' data it is the upper digit, according to the data in [3] it is the lower one.

White colour is not presented on the diagrams as it is not clear in which place of the sequence of colours it must be put, i.e. what the eyewitnesses mean reporting of "white colour". Two apparent variants are possible here: white-blue and yellow-white but there are no reasonable grounds to choose either. For the dependence between linear dimension and colour for "white" BL the average dimension 26,7 cm is found on the basis of 13 correlated events. For the dependence between lifetime and colour for "white" BL the average lifetime 12,3 s is found on the basis of 26 correlated events. If we impose a restriction of 20 m on the distance as well as on estimation of brightness, the position of the points will not change quantitatively but the number of descriptions corresponding to the searched correlations will be approximately a third less.

4. The correlation dependence between BL lifetime and characteristic linear dimension was investigated in [14]. It was shown that the points of dependence $t = t(d)$ (see fig. 10) found according to the methods described were well approximated by the expression:

$$t = t_0 \exp\left\{-\left(\frac{d_0}{d}\right)^K\right\}$$

The values of constants in [3] can be easily found by the method of the least squares: $K = 0{,}9$, $t_0 = 75$ s, $d_0 = 5{,}7$ cm. If we round off K to one than d_0 takes the value equal to 7 cm. The corresponding dependence is drawn in fig. 10 as a solid line.

The correlation $t = t(d)$ has been found on the basis of 256 descriptions when BL was observed from the moment of its formation up to the end of its existence. It is according to these 256 descriptions that BL lifetime t_i was estimated for each interval of values of its linear dimensions. Simultanuously for the same interval of values of BL linear dimensions, those for which the searched correlation took place were selected out of all 2082 descriptions and served the basis for finding the average BL observation time τ_i. In case τ_i appeared to be more than t_i , the average lifetime for the given interval of linear dimensions was taken equal to τ_i: $t_i \equiv \tau_i$. As a result of it the whole number of descriptions on the basis of which the correlation dependence had been found increased from 256 to 503.

5. The search of the correlation dependence between BL brightness and colour showed that the whole number of correlated events being equal to 310, the interdependence of these BL characteristics had a non-monotonous character illustrated by fig. 11 (for non-spectral colours correlation with brightness was stated in 89 descriptions).

It must be mentioned, however, that the solid curvs connecting the points in fig. 6 - fig. 11 are of purely illustrative character as they reflect the law of correlated property changes which the authors consider the most natural in the given position of "experimental" points.

6. The stated correlation laws of dependences of characteristic BL linear dimension on brightness and colour and those of BL lifetime on diameter seem to be obvious and need no special commentaries while the dependences of BL lifetime on brightness and colour as well as those of brightness on colour deserves attention because of their non-monotonous look. This could arouse a question of existence of various types of BLs if we took into account the data in [14] (dependences (3)). Unfortunately such a conclusion would be too hasty as the reliability of the correlation dependences presented is not high. The reason for this is the small amount of descriptions on the basis of which these

dependences were obtained. To check the results received it seems to be expedient to carry out the similar analysis on the basis of the essentially greater amount of the initial data.

It is interesting to note however, the coordination of dependences presented in fig. 9 and in fig. 11: BLs of blue and orange colours for which brightness is minimum have maximum lifetime. If we construct BL full energy dependence on colour on the basis of both correlation dependences discussed it will also have a non-monotonous look with two maxima in the fields of blue and yellow colours.

Then it must be noted that while searching for correlation dependences between BL lifetime, characteristic linear dimensions, brightness and colour only those facts were selected where BL was watched from the moment of its appearance up to the end of its existence.

It seems worth reminding that such properties as dimension, brightness and colour do not remain constant within BL lifetime. According to our data the changes of BL brightness during its existence were reported in 2,5% of events, the changes in colour- in 1,5% of cases and the changes of dimensions - in 4,1% of observations.

APPENDIX

1. The problem of reliability of the data reported by eyewitnesses of BL characteristics is considered rather important as till recently these reports are the only source to form a notion of this natural phenomenon. In the statistical analysis of BL properties represented in the given work as well as in the earlier ones [1-3,5,6] we have to deal with the BL descriptions made on the basis of the immediate impression as well as the information reported many years after the moment of observation by people of both sexes of various occupation and different age who naturally differ in educational level and the ability to evaluate spacio-temporal intervals as well as such emotional plastic characteristics as brightness and colour. Hence, for getting a valuable numerical characteristic of errors made during the analysis of information of BL properties reported by eyewitnesses we had to take into account as the ability of people to evaluate spacio-temporal intervals as the temporal stability of the data reported during the questioning. This is the subject of the parts below.

2. As many facts of comparison of BL dimensions with those of different objects of every-day life often occur in the descriptions, several groups of students were offered to evaluate diameters of various balls, oranges etc. in cm. by eye. The apprehension of the dimensions of objects offered happened to change within rather a wide range with maximum deviations up to 72%. The numerical data made by statistical methods revealed the average value for all the ranges of dimensions $\varepsilon \approx 0{,}034$, correct with confidential probability $P = 0{,}7$.

Approximately the similar situation appears with respect to temporal intervals. Those tested were presented either a picture or a description of an accident and then asked to evaluate the period of time during which the procedure had taken place (without any preliminary explanation of the essence of the test). In all the cases the temporal intervals from 5 to 25 c were taken. The average value of the given temporal intervals as it ought to have been expected appeared quite close to the real ones. At any rate the original value was included into the confident interval. but the value of the confident interval itself came to $\approx 18\%$ of the average value ($P = 0{,}7$) and individual divergencies twice from the original values were customary.

3. To evaluate the stability of data of BL properties 0,5 - 1 years after the first description the eyewitnesses were asked to fill in a form for computer processing. Then the data obtained by filling in the form were compared with the original descriptions. The results of the comparison revealed the fact that the observers were nor always sure in the reported BL characteristics. Quantitatively the value of such inaccuracy can be characterized by the value of divergencies between the initial reports and those obtained by filling in the forms.

If we assume for clarity the initial reports to be true and the data obtained by filling in the forms mistaken it is not difficult to find root-mean-square divergencies and relative errors throughout the total collected observation material for each range of BL diameter values and observation time. The results of this calculation are illustrated in table 15 where the relative errors $\mathcal{E}_\alpha$ and $\mathcal{E}_t$ are presented with confidential probability P =0,7 Near each value of an error in brackets there is a number of descriptions on the basis of which the values $\mathcal{E}_\alpha$ and $\mathcal{E}_t$ were obtained. From table 15 one can easily notice that the value of a relative error changes within wide limits while passing over from one range to another. But as the statistics used is not representative enough (the analysis of errors contains only 628 descriptions with respect to linear dimensions; 310 cases with respect to observation time), it seems reasonable to introduce an average relative error for all the characteristics throughout all the ranges. The average errors in charcteristic linear dimension and observation time will take the values: $\mathcal{E}_d$ = 0,14, $\mathcal{E}_t$ = 0,57 if we take into account the data of p. 2.

The results of such calculations for BL colour when each spectral component corresponds to the characteristic electromagnetic wave length are presented in table 16. The average value of a relative error for each colour is $\mathcal{E}_c$ = 0,025 (P = 0,7). The calculations have been made on the basis of 504 descriptions.

4. The least strictly evaluated BL characteristic i.e. brightness deserves special consideration.

In the reports of brightness of BL luminescence eyewitnesses used to compare it with lamps of different power, with a candle, a kerosene oil lamp, the fullmoon or they just stated "grimy", "faded", "bright", "semi-transparant", " transparant". Brightness characteristics of this kind are of course rather ambiguous. But we do not think it right to make the eyewitnesses to compare BL brightness only with that of electric lamps of different power: beginning with lamps of a few watts up to 500 w as it was made in the questioning in [3] . Indeed, as it has been stated above eyewitnesses were not quite sure in the accuracy of their reports and their evidence may be too far from real facts when they faced the necessity to chose a variant from some restricted set (including such exotic standards of luminiscence as lamps less than 10 w and greater than 150 w). However BL brightness comparison with electric lamps seems more convenient as it gives an opportunity to introduce numerical brightness characteristics. In this connection all the BL descriptions of our collected observation material were devided into 6 groups with brightness as a criterion (see п.4), with two extreme exotic ranges from [3] 200-500 w and greater than 500 w being united into one - greater than 200 w. The relative errors ε_b for each of 6 groups obtained according to the methods described above are presented in table 17 which is composed in such a way that our brightness standards can correspond to those accepted in [3] . For each brightness group the possibility for BL to come into it is given in brackets. The average meaning of relative error for all the groups except the first one will be equal to $\varepsilon_b = 0{,}3$.

References

1. Brand W. Der Kügellitz, Hamburg, Henri Grand, 1923.
2. Imjanitov I.M., Tikhy D.Ja. Beyond the Laws of Science. L., Gydrometeoisdat, 1980, p. 190.
3. Stakhanov I.P. Physical Nature of Ball Lightning. M., Atomisdat, 1979, 240 p.
4. James Dale Barry. Ball Lightning and Bead Lightning. Plenum Press, New York, London, 1980, 288 p.
5. McNally J.R. Preliminary report on Ball Lightning. - Second Annual Meeting of the Division of Plasma Physics of the American Physical Society, Gatlinburg, 1960, N 2-5, paper J.-15, p. 1-25.
6. Smirnov B.M. Properties and Nature of Ball Lightning. - Physics Reports, 1987, v. 152, N 4, p. 177-226.
7. Grigorjev A.I. Ball Lightning. Man and Element. L., Gydrometeoisdat, 1987, pp. 47-49.
8. Arago F. Meteorological Essays. London. London, Brown, Green and Longmans, 1855.
9. Matvejev L.T. Course of General Meteorology. Physics of Atmosphere. L., Gydrometeoisdat, 1984, p. 751.
10. Arabadji W.I., J. Geophys. Res., 1976, v. 81, N 36, pp. 6455-6456.
11. Grigorjev A.I., Grigorjeva I.D., Ognev A.M. On the Correlation of Thunder Activity in Atmosphere of the Earth with the Solar Activity According to the Results of Ball Lightning Observations. - Solar Data. Bulletin, 1985, N 5, pp. 91-93.
12. German J.R., Golberg R.A. The Sun, Weather and Climate. L., Gydrometeoisdat, 1981, p. 320.
13. Vitinsky Ju. I. Cyclic Character and Forecasts of Solar Activity. L., Nauka, p. 157.
14. Grigorjev A.I., Grigorjeva I.D. On the Correlation Dependence between the Lifetime and Characteristic Linear Dimension of Ball Lightning. - Meteorologija and Gydrologija, 1986, N 6, pp. 103-104.

15. Grigorjev A.I., Grigorjeva I.D. The Correlation Dependences between Ball Lightning Lifetime, Brightness and Colour. - Heads of the 111th All-Union Symposium on Atmospheric Electricity. Tartu, October 28-31, 1986. Tartu, Isdat. TGU, 1986, p. 228

16. Grigorjev A.I., Grigorjeva I.D. Meteorological Conditions Characteristic for Ball Lightning Observation. - Heads of the 111th All-Union Symposium on Atmospheric Electricity. Tartu, October 28-31, 1986, Tartu, Isdat. TGU, 1986, p. 229.

List of Tables

Table 1

Shape	Our data based on 2013 descriptions	Stakhanov's data [3] based on 878 descriptions	Rayle's data [3] based on 112 descriptions
a ball	0,9I22	0,897	0,875
an oval	0,0027	0,059	0,080
pear-shaped	0,0035	0,023	
spindle-shaped	0,0025		
ring-shaped	0,0045	0,002	0,027
a band			
shapeless	0,0I43	0,0I6	
disk-shaped	0,0079	0,00I	
a cilinder	0,0020	0,00I	

Table 2

Mixture of colours	white	rose	red	orange	yellow	green
0,0806	0,II29	0,0887	0,I29	0,I29	0,I533	0,0242

light-blue	blue	violet
0,I775	0,0645	0,0403

Table 3

	d_{max}	$\langle d \rangle$
our data	7,5	I8,5
Stakhanov's data	I2,5	22,5
McNally's data	I4,0	32,5
Raile's data	I2,0	45,4

Table 4

	our data, 1743 descriptions	Stakhanov's data, 928 descriptions	Raile's data, 98 descriptions	McNally's data
moved irregularly	0,086			
moved horisontally regularly	0,747	0,75	0,53	
fell down from the clouds	0,047			
rose up to the clouds	0,004			
rotated	0,05I		0,20	
went round the obstacle	0,026			
moved along the wires	0,039		0,I6	0,2

Table 5

<table>
<tr><th>Colour</th><th>Our data
1803 descrip-
tions</th><th>[3]
936 descrip-
tions</th><th>[5]
1467 descrip-
tions</th></tr>
<tr><td>white</td><td>0,I37</td><td>0,26I</td><td>0,23</td></tr>
<tr><td>red</td><td>0,I36</td><td>0,I92</td><td>0,I6</td></tr>
<tr><td>rose</td><td>0,029</td><td>-</td><td>-</td></tr>
<tr><td>orange</td><td>0,35I</td><td>0,I2I</td><td>0,I5</td></tr>
<tr><td>yellow</td><td>0,I7I</td><td>0,263</td><td>0,23</td></tr>
<tr><td>green</td><td>0,0I2</td><td>0,0I3</td><td>0,0I</td></tr>
<tr><td>light-blue</td><td>0,08I</td><td>0,088</td><td rowspan="3">0,I3</td></tr>
<tr><td>blue</td><td>0,027</td><td>0,0I6</td></tr>
<tr><td>violet</td><td>0,027</td><td>0,0I4</td></tr>
<tr><td>non-spectral
colours</td><td>0,037</td><td>0,032</td><td>0,09</td></tr>
</table>

Table 6

eyewitnesses' reports	Stakhanov's standards, W	our data 1321 descriptions	[3] 697 descriptions
a candle, a kerosene lamp	I0	0,067	0,092
"transparant", "dingy", "semi-transparant", a lamp 10-20 w	I0-20	0,078	0,I39
"dull", a lamp 50 w	20-50	0,I58	0,I83
"the moon", a lamp 100 w	50-I00	0,238	0,255
"bright", a lamp 150-200 w	I00-200	0,285	0,25I
"dazzling bright" a lamp > 200 w	200	0,I74	0,I00

Table 7

inside the house	50,I5 %
in the street of a settlement	24,50 %
in the field	9,53 %
in the forest	4,38 %
on the bank of a river, a lake	4,03 %
in the mountains	2,27 %
in the sky from the ground	4,I3 %
in the clouds from the plane	I,0I %

According to Raile's data [3] BL was observed indoors in 48 % of cases.

Table 8

<table>
<tr><th>BL appeared</th><th>our data 1924 des-criptions</th><th>[3] 1006 desc-riptions</th><th>McNally's 513 des-criptions</th></tr>
<tr><td>in thunderstorm</td><td>5I,00 %</td><td rowspan="4">70 %</td><td rowspan="4">73 %</td></tr>
<tr><td>together with a discharge of li-near lightning</td><td>I0,56 %</td></tr>
<tr><td>before thunderstorm</td><td>6,55 %</td></tr>
<tr><td>right after thunderstorm</td><td>8,78 %</td></tr>
<tr><td>in the rain</td><td>7,22 %</td><td rowspan="2">I7 %</td><td rowspan="2"></td></tr>
<tr><td>in cloudy weather</td><td>6,03 %</td></tr>
<tr><td>in fine weather</td><td>9,77 %</td><td>I3 %</td><td>0,006 %</td></tr>
</table>

According to Charman's data [3] BL appeared in thunderstorm with frequency $\approx$ 70%.

Table 9

BL appeared	share in %
in the channel of a linear lightning discharge	IO,I4 %
in the place of a lightning struck	II,I9 %
in the clouds	4,55 %
on a metal conductor	54,20 %
on a metal conductor together with a linear lightning discharge	7,34 %
on a metal conductor out of a spark	6,99 %
just a fact	5,59 %

Table 10

	I m	I-5 m	5-I0 m	I0-20 m	20-50 m	50-I00 m	I00 m
our data in % based on 1441 descriptions	35,04	33,04	6,05	6,59	6,42	4,27	9,53
[3] in % 965 descriptions	I6,40	34,30	I0,80	I0,60	I0,70	II,I0	6,20

Table 11

BL observation stopped when it	share in %
left the sight	40,69 %
exploded	25,86 %
retired into the ground	8,23 %
retired into a conductor	5,24 %
crumbled into sparks	6,45 %
faded quietly	I3,53 %

Table 12

month		I	2	3	4	5	6	7	8	9	I0	II	I2
BL number	units	20	9	I4	2I	II7	296	823	296	69	3I	6	II
	%	I,2	0,5	0,8	I,2	6,8	I7,3	48	I7,3	4	I,9	0,4	0,6

Table 13

time interval, h		0-3	3-5	6-9	9-I2	I2-I5	I5-I8	I8-2I	2I-24
BL number	units	48	I0	I29	449	392	229	I40	39
	%	3,3	0,7	9	3I,3	27,3	I6	9,7	2,7

Table 14

φ \ θ	25–40	40–45	45–50	50–55	55–60	60–65	65–70
20–50	0	I27	30I	39I	44I	I6	5
50–90	6	44	II4	200	II8	29	0
90–150	0	26	47	56	49	44	0

θ - latitude, φ - longitude

Table 15

range of values d in cm, t in sec.	0-I	I-3	3-5	5-I0	I0-I5	I5-20	20-25	25-30
$\mathcal{E}_d$ (N)	0 (3)	0 (I9)	0,I (36)	0 (I36)	0,08 (I00)	0,I5 (80)	0,I2 (69)	0,03 (8I)
$\mathcal{E}_t$ (N)	0,7 (72)	0,I (58)	0,I8 (33)	I,2 (35)	0,87 (I6)	I,05 (I4)	0,32 (7)	0,73 (I5)

30-35	35-40	40-45	45-50	50-55	55-60	60-70	70-80	80-I00	I00
0,II	0,07 (I8)	0,29 (3)	0,I4 (27)	0,I7 (4)	0,I8 (I0)	0,36 (3)	0,I9 (4)	0,I3 (I4)	0,I3 (I2)
0 (0)	I,5 (5)	0 (0)	0,26 (2)	0,07 (2)	0,25 (I7)	I,43 (3)	0 (2)	0,99 (2)	0,I (25)

Table 16

	Re	*Ro*	*O*	*Y*	*G*	*L-B*	*B*	*V*	non-spekt-ral co-lours
Colour *)	(0,I36) (0,I92)	(0,288) ( )	(0,35I) (0,I2I)	(0,I7I) (0,263)	(0,0I2) (0,0I3)	(0,08I) (0,088)	(0,023) (0,0I6)	(0,023) (0,0I4)	(0,064) (0,293)
$\mathcal{E}_c$	0,0II (58)	0,033 (I4)	0,005 (I70)	0,008 (80)	0 (0)	0,0I6 (52)	0,063 (I8)	0,065 (I7)	(I06)

*) near the mark of a colour in brackets colour probabilities in the total collection of descriptions are shown: the upper figure - according to the authors' data, the lower one - according to the data in [3] .

Table 17

standard of brightness	"a candle" "a kerosene lamp" (0,067)	"dull","trans-parant","semi-transparant" < 20 w (0,078)	"not bright" 20–50 w (0,158)	"the moon" 50–100 w (o,238)	"bright" 100–200 w (0,285)	"very bright" 200 w (0,174)
standard of bright-ness in [3]	IO W (0,092)	IO–20 W (0,I39)	20–50 W (0,I83)	50–IOO W (0,255)	IOO–200W (0,25I)	200 W (0,I)
$\mathcal{E}_c$	7,2 (I4)	0,6 (3I)	0,6 (27)	0,I5 (68)	0,08 (62)	0,07 (56)

List of Figures

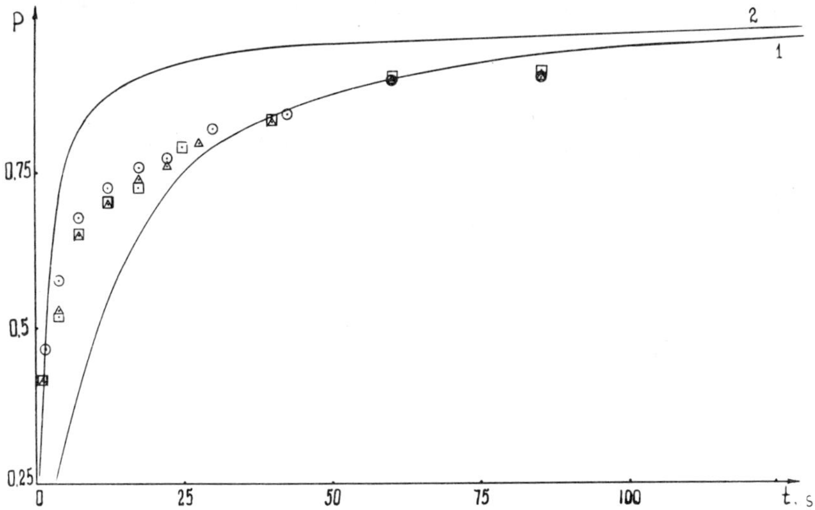

Fig. 1.

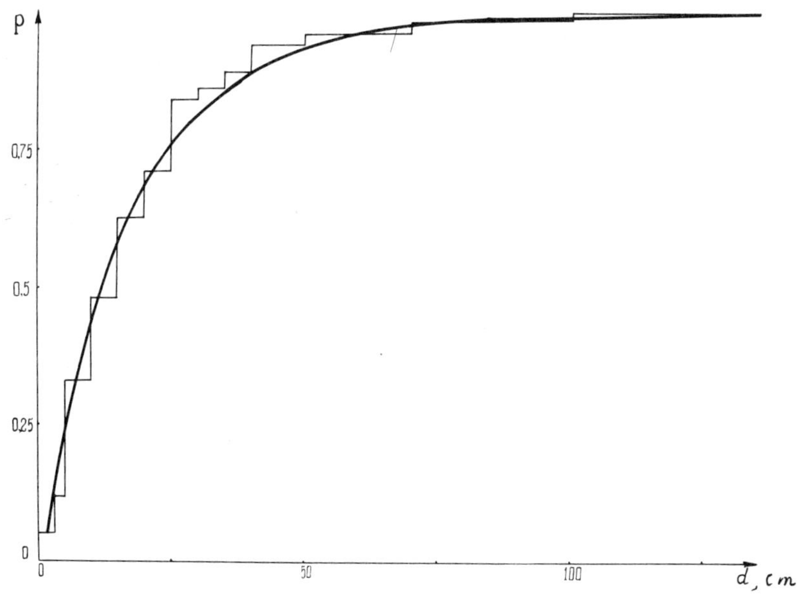

Fig. 2.

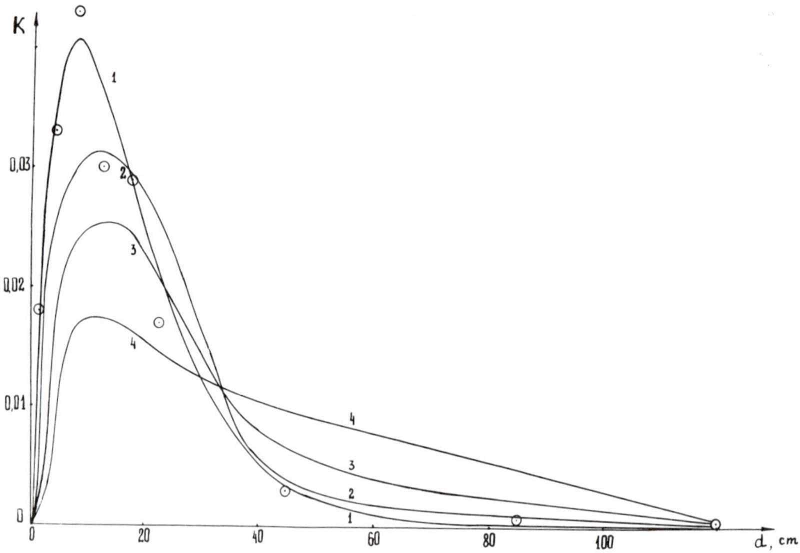
K
0,03
0,02
0,01
0
0
20
40
60
80
100
d, cm
1
2
3
4

Fig. 3.

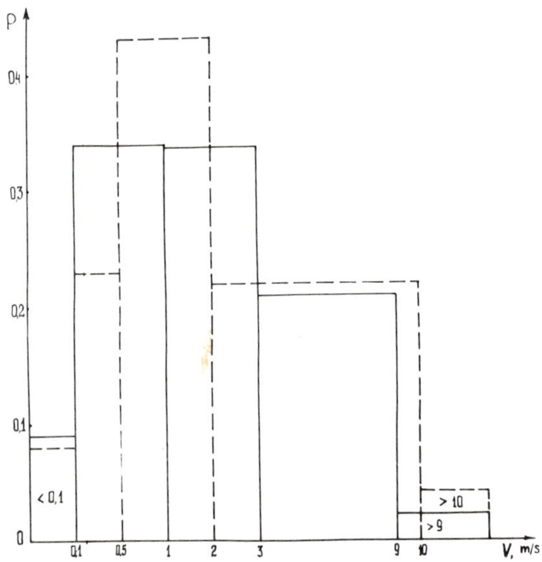
P
0,4
0,3
0,2
0,1
0
< 0,1
> 10
> 9
0,1
0,5
1
2
3
9
10
V, m/s

Fig. 4.

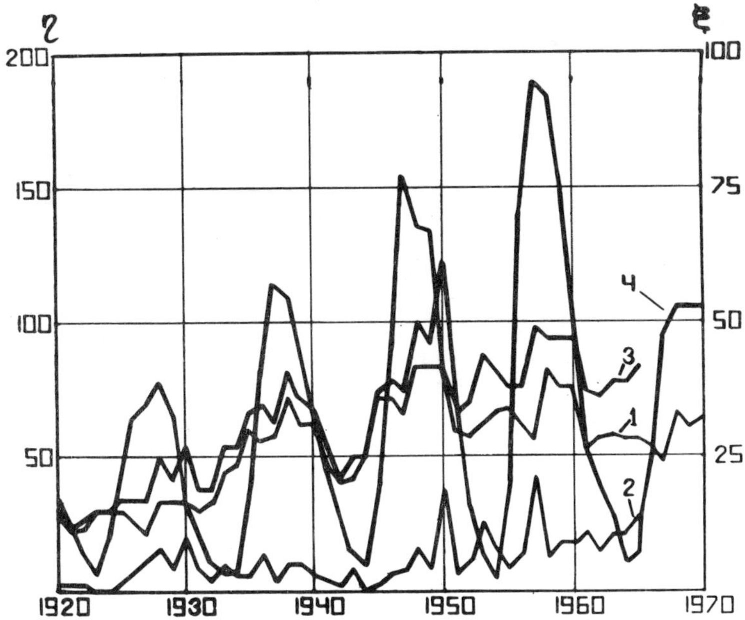

η
ξ
200
150
100
50
100
75
50
25
4
3
1
2
1920
1930
1940
1950
1960
1970

Fig. 5.

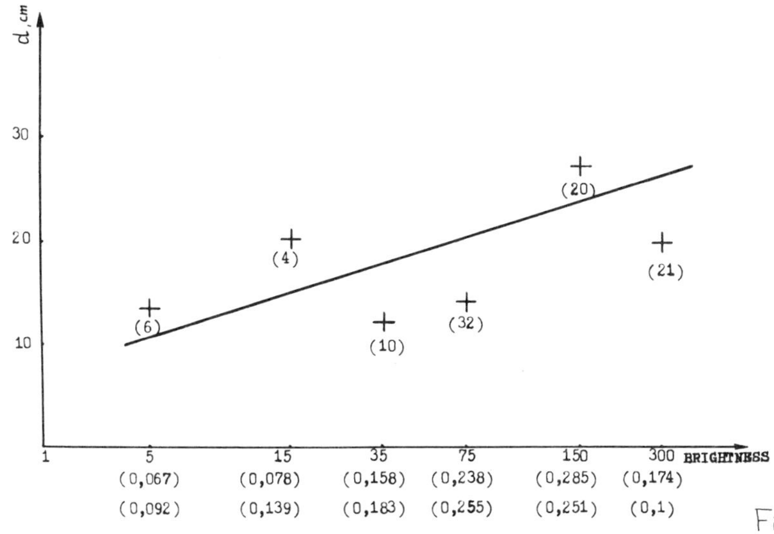

d, cm
30
20
10
(20)
(4)
(21)
(6)
(32)
(10)
1
5
15
35
75
150
300
BRIGHTNESS
(0,067)
(0,078)
(0,158)
(0,238)
(0,285)
(0,174)
(0,092)
(0,139)
(0,183)
(0,255)
(0,251)
(0,1)

Fig. 6.

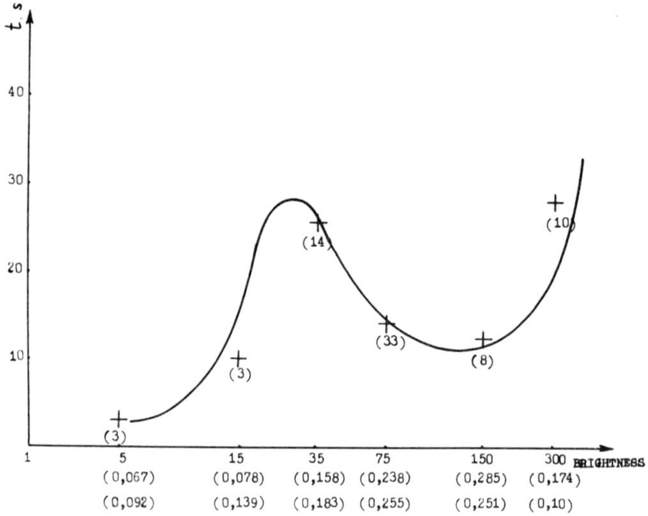
t,s
40
30
20
10
(3)
(3)
(14)
(33)
(8)
(10)
1
5
15
35
75
150
300
BRIGHTNESS
(0,067) (0,078) (0,158) (0,238) (0,285) (0,174)
(0,092) (0,139) (0,183) (0,255) (0,251) (0,10)

Fig. 7

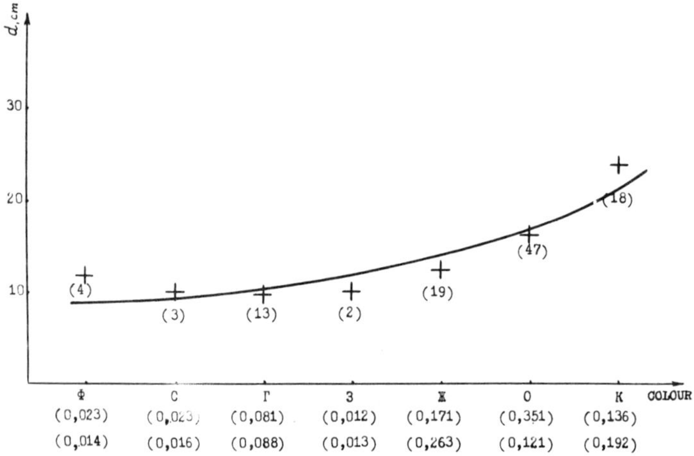
d,cm
30
20
10
(4)
(3)
(13)
(2)
(19)
(47)
(18)
Ф
С
Г
З
Ж
О
К
COLOUR
(0,023) (0,023) (0,081) (0,012) (0,171) (0,351) (0,136)
(0,014) (0,016) (0,088) (0,013) (0,263) (0,121) (0,192)

Fig. 8

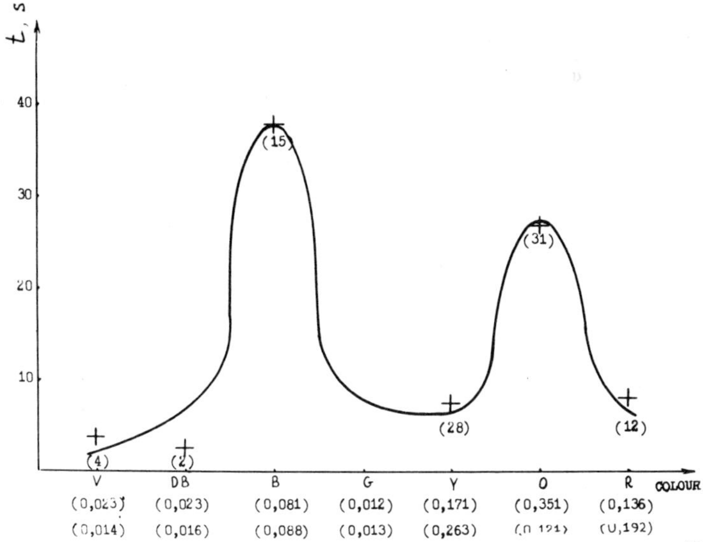
t, s
40
30
20
10
(15)
(31)
(28)
(12)
(4)
(2)
V
DB
B
G
Y
O
R
COLOUR
(0,023)
(0,023)
(0,081)
(0,012)
(0,171)
(0,351)
(0,136)
(0,014)
(0,016)
(0,088)
(0,013)
(0,263)
(0,192)

Fig. 9.

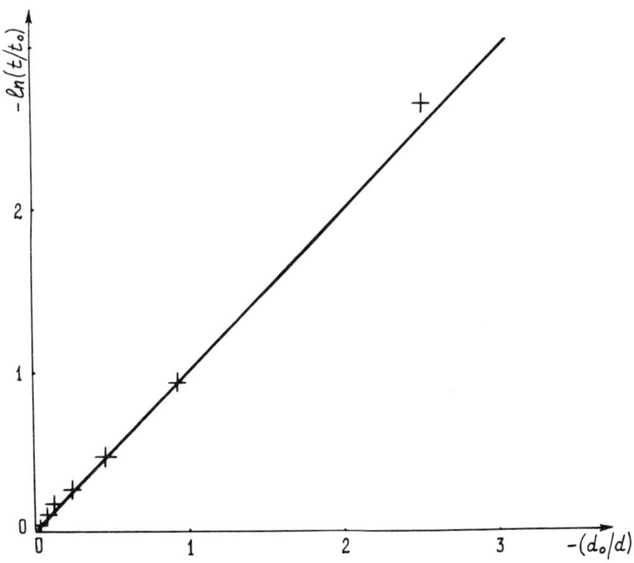
-ln(t/t₀)
2
1
0
0
1
2
3
-(d₀/d)

Fig. 10.

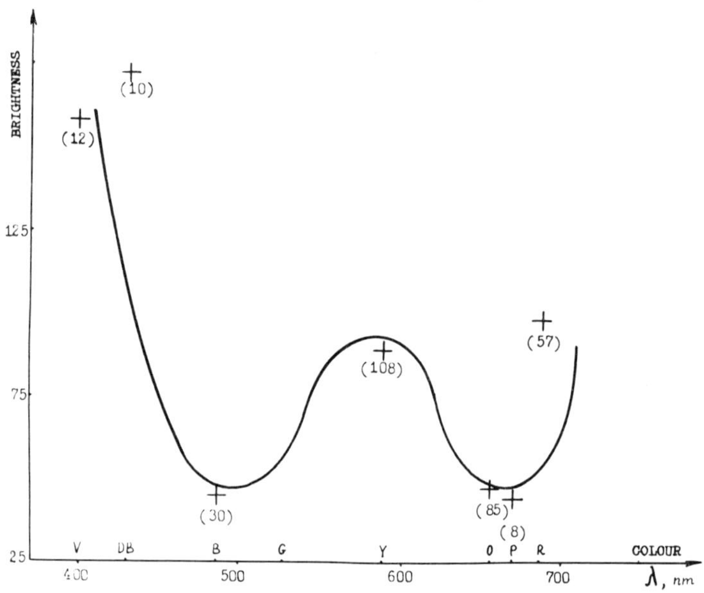
BRIGHTNESS
125
75
25
(12)
(10)
(30)
(108)
(85)
(8)
(57)
V
DB
B
G
Y
O
P
R
COLOUR
400
500
600
700
λ, nm

Fig. 11.

A FLUID-DYNAMICAL MODEL FOR BALL LIGHTNING AND BEAD LIGHTNING

Karl L.E. Nickel

Institut für Angewandte Mathematik
Universität Freiburg
Freiburg i.Br.
West Germany

PREFACE

The main purpose of this paper is to give an **explanation** for both **Ball Lightning** and **Bead Lightning** which does **not** need any "modern" science like *Plasma Physics, Magneto-Hydrodynamics* etc. but uses only "classical" science, namely ideas from **Fluid Dynamics**.

The proposed mechanism works **independent** of the physical or chemical status of the fireball, i.e.: if the ball is hot or cold; if it is filled with plasma, air, smoke, burnable gas or dust; and if electromagnetic forces are present or not.

With this ideas for the first time a formula for the **life time** of a Ball Lightning can be given. Also the **longitudinal instability** of a lightning channel can be explained which, most probably, is the origin of Bead Lightning. Hence, for the first time a formula for the **wave length** of this perturbance can be derived.

AMS MOS: Primary 76U05 Secondary 35Q10

This is an extended version of the :
CMA Research Report CMA-R11-86; February, 1986 at the
Centre for Mathematical Analysis of the
Australian National University in Canberra/Australia.

ABSTRACT

The phenomenon of Ball Lightning is not at all well-understood. There are even Scientists of the opinion that it does not exist. In the following paper, a new model for ball lightning is proposed. It consists of a combination of fluid (gas) flow and chemistry. It explains

(i) what ball lightning is;

(ii) how it develops from an ordinary streak of lightning; and

(iii) how it vanishes.

It is also shown that

(iv) there may be a strong connection between ball lightning and bead lightning; and that

(v) the proposed model is consistent with the observed properties of ball lightning.

1. Introduction

Ball Lightning and Bead Lightning have been observed for more than 2000 years. Hundreds of papers and many books deal with this subject. Most of the literature can be found in the two newest books on this subject: Singer: The Nature of Ball Lightning [8] and Barry: Ball Lightning and Bead Lightning [1]. They define these phonomena as follows:

"Ball lightning as well as bead lightning is considered by many to be an atmospheric electrical phenomenon observed during thunderstorm activity. It is reported to be a single, self-contained entity that is highly luminous, mobile, globular in form, and appears to behave independently of any external force. The ball lightning phenomenon is also known by many other terms: globular lightning, balls of fire, lightning balls, kugelblitz, globes de feu, tonnerre en boule, coup de foudre en boule, foudre globulaire, eclair en boule, fulmini globulaire, sharovoyi molnii, kugelynet, and klotblixtar, to name a few." Barry [1], p.3.

"Ball lightning is a luminous globe which occurs in the course of a thunderstorm. It is most often red, although varying colors including yellow, white, blue, and green have also been often reported for the glowing ball. The size varies widely, but a diameter of one-half foot is common. Its appearance is in striking contrast to ordinary lightning, for it often moves in a horizontal path near the earth at a low velocity. It may remain stationary momentarily or change course while in motion. Unlike the rapid flash of ordinary lightning, ball lightning exists for extended periods of time, several seconds or even minutes. Its path in many cases takes it indoors where the glowing ball occasionally passes quite close to an observer. It enters by a window or a chimney and may depart through such openings ...

In many cases witnesses note particularly that, although intensely bright, the ball gives off no heat and disappears silently. In others a great explosion occurs, displacing and damaging nearby objects.

Marked variation in these general characteristics is noted. The ball is seldom a true sphere; often it is a mass with only a generally defined form or with some protrusions. Sparks may be emitted by the body. The boundaries are sometimes well-defined, sometimes hazy and shrouded in mist. A hissing or crackling sound as in an eletrical discharge is often reported, or the ball may be silent. It falls directly from cloud to ground like a body of some mass, hovers, or even rebounds from the earth like an eleastic body. In some cases it seems to be borne by the wind; in others it travels directly opposite the prevailing wind direction." Singer [8], p.2/3.

In the final chapter of his book, Stanley Singer writes ([8], p.146):

> *"Despite reports of upwards of one thousand observations in the literature and more than a half dozen comprehensive, detailed reviews of the problem, including two monograph volumes, published in the last 125 years, ball lightning remains one of the greatest mysteries of thunderstorm activity. Still less can it be said that experiment has succeeded where theory fails in duplicating more than the simplest appearance of the glowing spheres. A great number of theories have been offered, themselves involving a bewildering variety of phenomena in widely separated areas of science."*

It is the aim of the following paper to add one more theory to that large number. It is not so much an entirely <u>new</u> theory as a combination of at least three theories which have been developed independently from each other and which have been discussed in the literature for some considerable time. The author hopes that by this process a new understanding can be reached of these mysterious phenomena:

2. Ball Lightning

2.1 Observed Properties

Over the centuries a very large number of observations of ball lightning has been made. The recorded characteristics vary considerably and often contradict each other. In this Sub-Section, therefore, no distinction is drawn between the different properties. All the data provided are taken from the books of Singer [8] and Barry [1]; in particular from Section 3.2 of Barry's book, [1], p.35 to 43.

Shape:	Mostly spherical; also oval, teardrop, rod, ...
Size:	Mostly 10-40 cm; down to some cm; up to several metres.
Color:	Mostly red to yellow; also white, green, bluish, purple...
Structure:	Solid appearance; or rotating structure; or burning appearance.
Motion:	Sometimes motionless; can move, even against the wind; speed in general not high (some m/s).
Sound:	Mostly soundless; sometimes a hissing or crackling noise.
Odor:	Sometimes sharp and repugnant; but not always apparent.
Heat:	Sometimes no heat detected; sometimes high heat emission causing burns at things, animals and people.
Lifetime:	A few seconds, lasting up to minutes.
Decay:	Silent decay mode or explosive mode.
Lightning Dependence:	Often associated with a proceeding ordinary lightning discharge.
Attraction to Enclosure:	Yes; ball lightning entering a building by way of a chimney or through a partially open window or door are observed.
Altitude Dependence:	Predominantly observed at low altitudes.
Stability:	Very stable against atmospheric turbulence; ball lightning "bouncing off" the ground or walls.
Electrical Charges:	Sometimes not charged; often charged with high voltage producing an electrical shock upon touching.

2.2 Previous Explanations

There is by now a very large number of "explanations", "models" or "theories" to the problem of the existence of ball lightning. It is impossible to cite them all in this paper. See for references the books of Singer [8] and of Barry [1]. It is, however, necessary to describe some classes of theories and discuss their basic ideas and their shortcomings. They are examined in the following sub-sections:

2.2.1 The "all-electricity" models

It is widely assumed, ...

> *"that ball lightning is associated with the ordinary lightning discharge and is an electrical phenomenon. This association is supported by reports that describe a ball lightning appearing simultaneously with a nearby ordinary lightning discharge. About 90% of the ball lightning observations reported occurred during thunderstorm activity."* [1], p.40.

These observations suggest immediately a close connection between these two types of lightnings. Singer [8], p.80 writes about this:

> *"The reappearance of theories after long intervals of time, as much as a century, has not been unusual in studies of ball lightning. The view that ball lightning is composed of the substance of ordinary lightning and is formed by the separation of a portion of an ordinary flash of lightning has been repeated in this way:"*

In other words, to the naive observer a ball lightning would be "a spherical bag, filled with electricity, which is the remainder of a lightning stroke". How can one define a sound theory out of this?

During the centuries many "all-electricity" models have been proposed. Their basic idea is always, that only one mechanism is involved. This mechanism is thought to be of a purely electrical nature, e.g. interaction of electric and magnetic fields or standing waves with plasma or highly ionised atoms, molecules or particles. Up to now all such onesided theories have failed to explain all the known properties of ball lightning. They cannot, for example, explain the fact that sometimes

> *"... ball lightning has been observed to come into contact with conductors without being immediately terminated ..."* [1], *p.43.*

2.2.2 The "chemical" models

The shortcomings of the previous models together with some observations (ball lightning sometimes spout flames, burn, explode ...) suggest a chemical process is involved.

> *"According to this theory ball lightning is a secondary phenomenon initiated by ordinary lightning and possibly maintained by the reaction of chemical compounds also produced by the previous flash. This view has been accepted by some authorities."* [8], *p.87.*

There is, however, no reason why such a chemically reacting gas mass should organize itself into a globular structure. Also, such a gas mass could certainly emit heat, light and the odor of burnt gases and could finally explode. But how could one explain motion, even against the wind, bouncing off the ground, etc.? Hence, unfortunately, the purely chemical models must also be discarded, even if one combines them together with electrical models.

2.2.3 The "vortex" models

Some observations speak of internal motions in ball lightning, similar to that in a smoke ring. The earliest such reference seems to be Lucretius (cited after Singer [8], p.11). Therefore, again and again such "vortex" models have been proposed, consisting of rapidly spinning plasma or ionized gas and thus producing magnetic and electric fields which keep everything together (see Singer [8], p.94 to 133). To the knowledge of the author all such models, are thus never purely hydrodynamic structures but assume always to be in combination with electromagnetical forces. Hence, why are they not terminated when they become grounded? (see [1], p.43).

2.3 The Main Problem

It is not easy to form a consistent theory which explains all (or, at least, most) of the observed characteristics of ball lightning. Singer writes [8], p.3:

> *"The great variation in properties reported has been a source of much confusion in providing a clear explanation of ball lightning. Theories have been unusually numerous. Most explanations accept the role of storm electricity at least in initiating the glowing mass. The continuing activity of the ball may then be ascribed primarily to chemical reaction or to electrodynamic processes. Chemical theories presented historically include suggestions that nitrogen triiodide, hydrogen and oxygen mixtures, or ozone formed by storm discharges provide the substance of the ball and the energy released in subsequent decomposition. Active nitrogen produced by a lightning stroke and a nitrogen discharge "flame" in the atmosphere resulting in nitrogen oxides have been proposed. Wholly electrical theories have presented the ball as a brush discharge. The separation of a short segment of an ordinary stroke of lightning as a vortex has been suggested. Vaporization of a metal such as copper by an intense lightning stroke may produce the ball. Electrically charged dust particles, raindrops, or ions of the atmospheric gases in assemblies in which neutralization of opposite charges by recombination is somehow slowed have been discussed. Plasma theories provide many of the most recent models considered, based on the rapidly increasing knowledge of matter at high temperatures from thermonuclear research."*

If one considers all the observations of ball lightning made over the centuries one can pose the following:

MAIN PROBLEM FOR BALL LIGHTNING

HOW TO EXPLAIN A GLOBULAR AIR MASS WHICH IS (IN MANY CASES) HOT AND BRILLIANT AND PRODUCES ELECTRICAL SENSATIONS?

One may break up this (complicated) problem into three different and (possibly) unconnected
Partial Problems:

i) Is there a mechanism which keeps together for seconds or even minutes a globular mass of air, gas or plasma regardless of its temperature, chemical or electrical state?

ii) Is there a (possibly different) mechanism which explains how such a mass may have the observed electrical properties?

iii) Is there a (maybe again different) mechanism which produces heat and light in such an air mass?

It will be shown in the next Section 2.4 that satisfactory answers to all of these independent questions can easily be found. They are, surprisingly enough, already all contained in the literature on ball lightning. Hence, the only originality of the following section lies, perhaps, in the idea to bring these different explanations together.

2.4 The New Model

The proposed new model for ball lightning is composed of these independent mechanisms which - taken together - explain the main known properties of ball lightning. For the convenience of the reader these three partial problems which were stated in the previous Section 2.3 will be repeated here.

2.4.1 First Partial Problem

> *Is there a mechanism which keeps together for seconds or even minutes a globular mass of air, gas or plasma regardless of its temperature, chemical or electrical status?*

A simple answer to this first question has been known for nearly a century. It consists of the hydrodynamic forces which are produced by a globular vortex field of constant vorticity. It is known as the "Hill vortex" (see for example Lamb [5]).

Consider the steady motion of an incompressible fluid in the three dimensional x,y,z-space.

Define $r \geq 0$ by $r^2 := x^2 + y^2 + z^2$

and $\varrho \geq 0$ by $\varrho^2 = y^2 + z^2$.

Let $a > 0$ and $U > 0$ be given real constants. Define the stream function $\psi(x,y,z) \mathrel{\hat{=}} \psi(\varrho,r)$ by

$$(1) \quad \psi(\varrho,r) := \begin{cases} \frac{3}{4} U\varrho^2 (1 - \frac{r^2}{a^2}) & \text{for } r \leq a \\ -\frac{1}{2} U\varrho^2 (1 - \frac{a^3}{r^3}) & \text{for } r \geq a. \end{cases}$$

The part outside the sphere $r = a$ describes the fluid flow produced by a parallel stream of velocity U in the x-direction and by a dipole which is located at the origin. The inner part gives a fluid without discontinuity at the surface of the ball $r = a$. The velocity components u in the x-direction and v in the ϱ-direction can be evaluated from ψ by the well known expressions

$$u = -\frac{1}{\varrho}\frac{\partial\psi}{\partial y}, \quad v = \frac{1}{\varrho}\frac{\partial\psi}{\partial x}.$$

From (1) it follows that ψ vanishes for $r = a$ and for $\varrho = 0$. Hence $y = z = 0$ and $x^2 + y^2 + z^2 = a^2$ are streamlines (with the value $\psi = 0$ for the stream function). Hence (1) describes the fluid flow around a sphere of radius a, with the fluid moving at infinity with the constant velocity $U > 0$ in the positive x-direction. Some of the streamlines of this flow in the x,y-plane for $z = 0$ are sketched in Figure 1, which is taken from the book of Lamb [5], p.246.

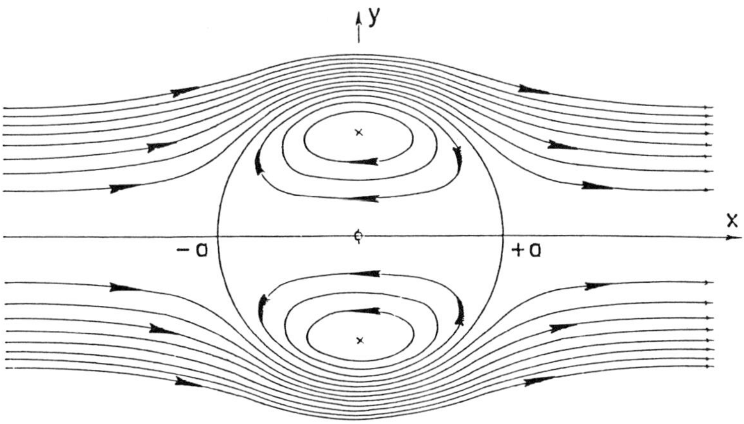

Figure 1: Fluid flow around a sphere. The streamlines in the x,y-plane are drawn for equidistant values of ψ. From Lamb [5], p.246 (arrows added).

Figure 1 shows quite clearly that the fluid flow is separated in two disjoint regions, i.e., inside and outside the sphere with diameter $r = 0$. From an "outside" point of view the fluid flows around a solid ball. If one looks "inside", however, one has a moving fluid mass which is trapped inside a globular volume. The two fluid masses are completely separated from each other and do not mix at all. This fact will be quite important for the later explanation of the ball lightning phenomenon.

For many decades Hill's vortex was the only existing solution of such a moving fluid mass separated from the outside. During the last few years, however, it has been proved that there exist infinitely many such solutions of Eulers equation. See the papers of Fraenkel and Berger [3] and Norbury [6], where more references can be found. The shapes of the fluid masses are all "discus-shaped" or even toroidal with the ball of Hill's vortex as one extreme solution. It would be quite interesting to know if "cigar-shaped" structures are possible, too.

The Hill vortex has one big disadvantage: the gradient of the velocities at the ball surface is not continuous. Hence there is a jump in the stress at these points. Therefore such a steady fluid flow cannot exist, it has to decay due to the viscous forces.

While, then, the Hill vortex cannot be the solution to the real problem, it can at least serve as an approximation to an existing unsteady solution of the Navier-Stokes equation which form a ball-like mass of moving fluid.

The streamline pattern of Figure 1 is obviously quite similar to that of a smoke ring. In Figure 2 a photograph of such a smoke ring is presented. This photo was taken from the wonderful book of van Dyke [2].

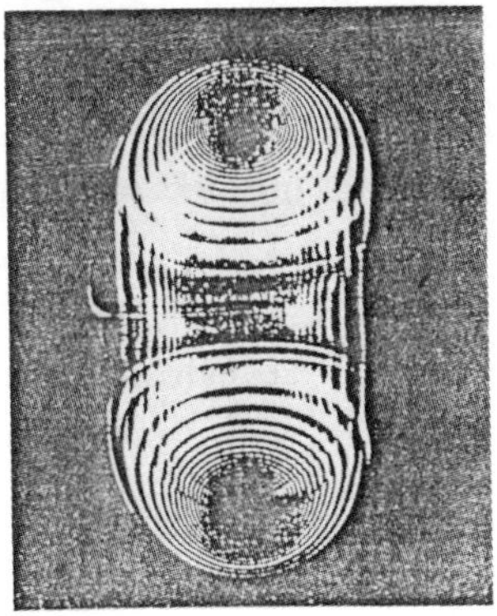

Figure 2: Flow pattern inside a smoke ring. From van Dyke [2], p.44.

From this similarity one expects that both fluid phenomena (Hill's vortex and smoke ring) behave similarly.

During the last decades smoke rings have been studied in great detail. But to the knowledge of the author up to now no theoretical nor computational model of such a fluid flow exists which is completely satisfying, e.g., which can model the exact decay rate of a smoke ring (or of a Hill vortex). Things are even more complicated since it has been shown that smoke rings can exist in two states, namely in laminar and in turbulent form. The latter is quite clearly not yet ready for a theoretical treatment. Hence one has to rely on experiments. They show some facts which most smokers certainly know:

(i) Smoke rings are quite stable configurations.

(ii) They last for up to some minutes.

(iii) They travel on their own for some distances, even against the wind.

(iv) They bounce off the ground or a wall.

(v) They are attracted by openings.

(vi) They contain a certain air mass which, during their lifetime, is (nearly) <u>not</u> mixed with the surrounding air.

(vii) While they travel, a small trace of smoke is (in contradiction to (vi)) left behind them.

BALL LIGHTNING REPORTEDLY POSSESSES ALL THESE PROPERTIES. BY ANALOGY ONE MAY, THEREFORE, CONCLUDE THAT THE PHENOMENON OF BALL LIGHTNING IS IN PART SIMILAR TO THAT OF SMOKE RINGS (OR TO THAT OF A HILL VORTEX).

It is the belief of the author that this statement is true and that a ball lightning exists primarily as a rotating air or gas mass similar to that of a Hill vortex or of a smoke ring. This would explain why different ball lightning have very similar features even if they are otherwise very different.

Ball lightning has the "mechanical" properties (i) to (vii) listed above for smoke rings. These characteristics have been observed for a ball lightning whether it was hot or cold, it radiated bright light or not, it was electrically charged or not. Hence, it seems fair to assume that there is also a "mechanical" explanation behind these "mechanical" features and that this is in fact a hydromechanical force.

2.4.2 Second Partial Problem

Is there a mechanism which explains how such a mass may have the observed electrical properties?

The answer to this question is obviously quite simple: By the hydro-dynamical mechanism explained in the previous section the air mass inside the ball is completely separated from the air outside of it. If, for instance, the air trapped inside is initially ionized then it remains in that state for a certain (theoretically infinite) time since there is (nearly) no interchange between the two air masses. Hence, with respect to the outside, this ball can be regarded as a solid mass which signals its presence only through its conductive surface. It resembles, therefore, a ball capacitor with the usual capacity, which is independent (!) of the material, i.e., of the degree of ionization of the enclosed air mass. The electric charge of this capacitor depends, therefore, only on its surface (its diameter) and on the initial voltage. The latter can be assumed to come from an ordinary lightning stroke which precedes or triggers the ball lightning. It would then have the order of magnitude of 10^6 volt or higher.

Hence, it becomes quite clear that such a large electric charge (against ground) may give terrible electrical shocks to humans or animals, which could knock them unconscious or may even kill them. It could produce an overload in electrical wires or even shortcircuit them. Due to the high voltage of the electrical charges of such a ball capacitor one would also expect sometimes ionization of the air around it and the production of St. Elmo's fire. If it comes close to the ground or to some grounded mass it may also emit sparks, or small lightning flashes.

All these - reported - characteristics can, therefore, be explained quite naturally once a closed ionized air mass can be kept stable over some time interval.

It is quite easy to give experimental evidence that electrical charges can be carried for some distance by an enclosed air mass. Such experiments were performed by the author several decades ago. The set up is sketched in Figure 3. Here smoke rings were produced in the usual way from a smoke filled box with a circular opening on one side by tapping the elastic back at the other side. In front of the hole a brush discharge was produced which ionized the smoke-air mixture inside the ring and gave it an electrical charge at the same time. The existence of this charge was then verified at a distance of some meters by an electroscope touching the smoke ring. One may, alternatively, touch the vortex ring with a finger - which is quite convincing!

This experiment is very simple. The smoke is, of course, not necessary. It is only used to show the position of the - otherwise invisible - vortex ring. It may be noted, however, that the most difficult problem in this experiment was ... hitting the electroscope with the smoke ring. Such vortex rings never go in a completely straight line but have the tendency to wander away from the predicted path.

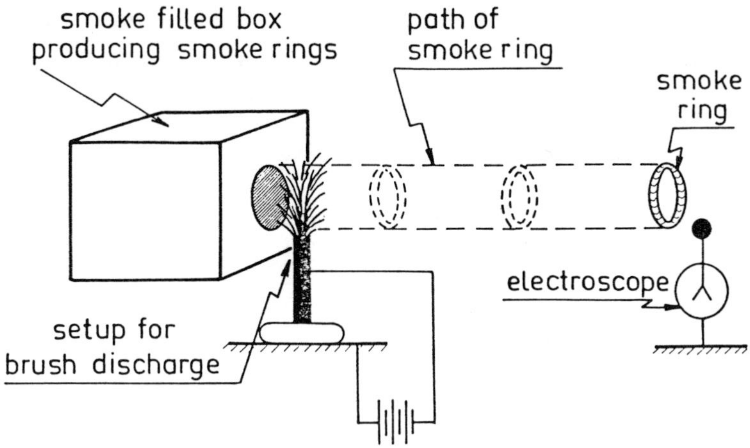

Figure 3: Schematic sketch of an experiment which shows that electrical charges can be carried by a smoke ring over some distance.

2.4.3 Third Partial Problem

Is there a mechanism which produces heat and light in such an air mass?

This last question is actually the easiest to answer. By searching through the literature one finds every detail of it already mentioned. It is either

(i) *"a localized burning process requiring combustible gases; or*

(ii) *a localized heating process requiring vaporized materials; or*

(iii) *a localized excitation of air molecules to metastable states leading to energy storage."*

See Barry [1], p.195.

The only problem which has never been solved in any of the previous publications is the question of how a hot air mass can possibly be confined in an isolated configuration. But the answer to this has already been given in Section 2.4.1.

For simplicity only the first possibility will be treated, i.e., the burning of combustible gases. This will lead, however, to two different mechanisms of burning and, consequently, to two different characteristics of ball lightning according to the amount of fuel which is contained in the lightning ball:

2.4.3.1 Low-density combustion

Experiments were conducted by Nauer, Behr, Barry and others which showed, that a very low-density combustion of hydrocarbon (or other) gases is possible if triggered by an electric spark (see [1], p.164 to 176, especially p.169 for reference). In these experiments the hydrocarbon volume percentage was much lower than that necessary for ordinary combustion ([1], p.170). The burning process was described as a "cool flame". Normally, it decayed silently.

A ball lightning with such an interior combustion would, therefore, be of the silent decaying type with no (or only little) heat emitted and it would have a low luminosity.

2.4.3.2 High-density combustion

If the gas content in the inside of a ball lightning is initially very high (e.g. pure methane, no air at all), a very different process is to be expected. Since there is no oxygen inside the ball, no interior combustion is possible. At the boundary between the inside and outside flow, however, things will change. Due to the diffusion of oxygen molecules to the inside and, possibly, turbulent mixing in the case of a turbulent ball lightning, a thin layer of flames will be nourished by the inside gases. Hence, in this case a ball lightning will appear as a "ball on fire". Eventually, even an outburst of flames can be expected as well as a "crackling noise" and the emission of fumes and smoke. This phenomenon will last as long as there is enough pure gas left inside the ball. But eventually the oxygen from the outside penetrates deep inside the ball and produces there an explosive mixture: Hence, the explosive decay will be just the "natural" end to such a mechanism.

All these considerations support, therefore, the opinion of the author that there are two different kinds of ball lightning, both having the same mechanical origin: One type has a cool flame and an interval combustion, silently running out of fuel. The other type burns with a hot flame - but only on the outside of the ball - and ends its life with an explosion. One could even hypothesize that these two different types are associated with the difference between laminar and turbulent fluid flow.

It should, finally, be emphasised that such a burning process is not always necessary in a ball lightning. Evidence from observations shows that in most cases a ball lightning is emitting heat or, at least, light. But if at the initial stage no burning fuel is available a ball lightning could still be visible if filled with dust, smoke or small particles. Some observations suggest that this type of not-burning ball lightning occurs on occasions.

2.5 The Birth of a Ball Lightning

It is the belief of the author that there may be several different mechanisms which are able to produce a ball lightning as described in the previous Section 2.4. One such possible mechanism will be sketched in this section. A second will be presented in the next paragraph in Section 3.3, where the birth of a bead lightning will be considered.

There are very few cases reported, where observers have actually <u>seen</u> the birth of a ball lightning, i.e., the development of such a phenomenon from a preceding lightning stroke. One of them, which is neither included in the book of Singer [8] nor in that of Barry [1] is the report of W. Hassenstein [4]. A shortened excerpt, translated from German, is as follows:

> *"It was in August of 1938. Mr. Hassenstein was driven by a driver in a closed car on a broad highway leaving Berlin. There was a heavy thunderstorm. The road was empty except for one car in front of them approximately 200 to 300m away. A lightning struck this car. The car was not affected. The lightning stroke glided down the back of that car, hit the ground and <u>curled up</u> there to form a ball of approximately 1 metre diameter. It then rolled at a steady speed of approximately 4m/sec toward them, exactly in the middle of the road. There was plenty of time to avoid the ball but the driver did not change lanes (he probably did fear to show fear). The ball then was overrolled by the car. There was a stink like burned rubber. But the ball was <u>not</u> affected. It reappeared unchanged behind the car and rolled further with the same size, same speed, and always in the middle of the road. About 300m behind the car it disintegrated silently without an explosion.*

A careful inspection of the motor and the electrical cables afterward showed no sign of any damage.

Both Mr. Hassenstein and his driver reported the same observations.

It was later learned that under the middle of the highway a high tension cable was buried. Was this the reason why the ball lightning always kept exactle to the middle of the highway?"

If one wants to explain this incident one first has to look at the initiating lightning stroke. What is such a stroke? Most people immediately associate the following sensations with a lightning stroke: blinding light, deafening thunder, high voltage (up to 10^9 volt), high current (up to 10^5 Ampere), ionized hot air or plasma (up to 30 000° Kelvin) in a small channel (up to 10cm in diameter). It is seldom mentioned, however, that there is also a strong <u>jet of air</u> produced by any lightning stroke. The reason for that production is quite simple: If a lightning stroke goes, say, from the ground to a cloud then this means that the (negative) electrons flow upward in the lightning channel. Hence the positive charged ions in that channel are forcibly driven down toward the ground thus producing a very strong downward jet. The velocity of that jet is of the order of 10 m/s to 100 m/s (some experts think that it may even reach supersonic speed and thus be partially responsible for the thunder). The jet produced by such a typical "return stroke" is sketched in Figure 4.

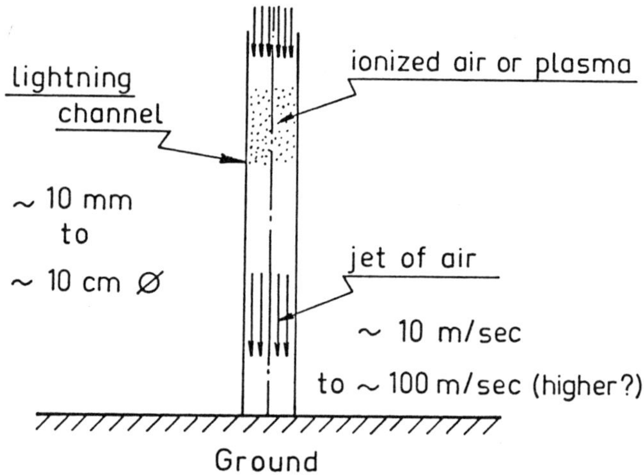

Figure 4: Sketch of a typical "return stroke" producing a jet of ionized air.

As soon as this jet of air hits the ground it can not proceed. One may therefore expect it to "roll up", as sketched in Figure 5.

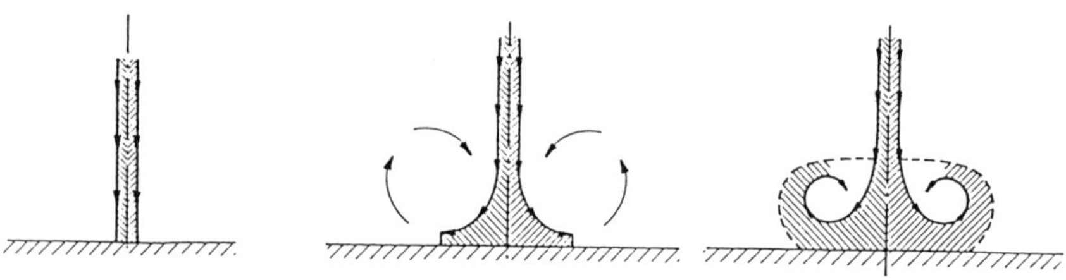

Figure 5: Sketch of a downward jet hitting the ground and subsequently rolling up to form something like a Hill vortex.

This roll up sketched in Figure 5 is produced only by aerodynamic forces. <u>No</u> electromagnetical "pinch effect" is necessary to form this phenomenon.

The report of W. Hassenstein [4] may be an observation of a ball lightning which was generated by a mechanism like that sketched in Figure 5. The rest of his story can also easily be explained: It goes without saying that an ionized spherical gas mass produced in such a way would most certainly receive an initial electrical charge making it act as a ball capacitor (see Section 2.4.2). Hence this ball lightning would be attracted by the field of the high tension cable. The chemical "fuel" responsible for its brightness could have two sources: Firstly, the hot lightning stroke on the road could produce hydrocarbons evaporating from the hot bitumen. Secondly, some fuel could also come from the two cars involved: In those days the internal combustion in a gasoline motor was far from perfect. Hence, the exhaust gases contained still a considerable amount of unburnt hydrocarbons. Therefore the ball lightning could have fed on the exhaust gases of the two cars and remained for a completely different reason always exactly in the middle of the road in the wake of the cars.

As stated in the beginning most probably many different mechanisms could produce a ball lightning. The previous explanation of one observed event does not mean that all ball lightning are produced in the same way. Other possibilities involving a preceding lightning stroke will not be considered here (see, however, the explanation for the formation of bead lightning in Section 3.3). It may even be that such an aerodynamic phenomenon evolves without any accompanying lightning flash, e.g., produced by a jet of air coming out of an underground opening, without bearing an electrical charge and lacking any combustion process.

There are some (rare) observations of a ball lightning appearing within a closed room. In this paper no explanation will be given to these reports.

2.6 The Death of Ball Lightning

If one accepts the model of ball lightning in Section 2.4 then there are 2 possibilities which will terminate such a phenomenon:

2.6.1 Aerodynamical Decay

After a life span of seconds or few minutes the kinetic energy of the whirling gas mass is eaten up by friction forces. This is the same as in a "dying" smoke ring: The ball will expand and dissolve itself in the surrounding air. The possibility of a chemical reaction depends on the gas content inside the sphere prior to its dissolution:

(i) If there is not much combustible gas left then the ball lightning will just fade away.

(ii) If there is a medium amount of combustible gas then there will be a rapidly expanding hot flame like a torch.

(iii) If, however, a large amount of unburnt gas is left then an explosive mixture with the oxygen of the air may create a huge explosion.

2.6.2 Chemical Decay

This mechanism has already been described in Section 2.4.3. Depending on the concentration of fuel gas inside there are the two possibilities of

(i) Silent decay; or of

(ii) Explosive decay

All these different types of decay have been recorded in observations.

2.7 An Example

The following event has not been included in the list of references in the books of Singer [8] or Barry [1]. An excerpt has been taken from the report of J. Schneider [7] and translated into English:

> *"On the 6th of August of 1938 many terrible thunderstorms closed together at the mountain "Wasserkuppe" in the Rhön-montains in Germany. Two men were standing together on top of the "Pferdskopf". One of them describes that a ball the size of a football (30cm diameter) of bluish color was brushing past the legs of his friends. He sensed a smell like something smoldering. The ball then moved toward his leg. He wanted to touch it because he was curious what it was. He then got a terrible electric shock and had the feeling as if his leg was torn out. He was knocked unconscious. When he regained consciousness much later in an ambulance station nothing wrong could be found with him or his trousers."*

This report - as so many others - is quite interesting both in what it tells and what is omitted: The ball lightning had the usual shape (spherical) and size (30cm). The color (bluish) is not very often recorded. Was the ball light emitting and brilliant? Probably, not, otherwise it would have been recorded. The structure had a solid appearance and the ball was moving on its own. No sound was observed, but a strong odor is recorded. Obviously, also no heat was emitted. Nothing was said about the decay. Also there is no mention of a preceding ordinary lightning (but there were flashes in abundance that afternoon). The ball seemed to be attracted to the legs of the men; also the occurrence was very close to the ground. The ball was obviously quite stable and had a formidable electrical charge. Hence, nearly all the typical features of Section 2.1 were observed.

Explanation: The ball was probably formed by a lightning stroke to the "Pferdskopf" with a low content of burning gas inside (there is a quite wet pasture nearby where marsh gas may have developed). This accounts for the odor and the absence of flames. By the solid appearance of the ball a laminar motion of the gas flow is suggested. The ball lightning obviously had a quite high electrical charge on it which certainly came from the preceding lightning stroke. The decay was then probably silent since the friend did not mention an explosion.

3. Bead Lightning

3.1 Observations

The following descriptions, observations and remarks have all been taken from Barry [1]:

"Bead lightning is also known by the terms pearl lightning, chain lightning, Perlschnurblitz, and clair en chapelet, to name a few. The bead lightning form is fairly well documented and discussed in the scientific literature. Although the number of reports and discussions of bead lightning is decidely less than of those involving ball lightning, bead lightning is an accepted form of atmospheric lightning.

Bead lightning has been reported most commonly to occur from one cloud to another, forming a discontinuous line of luminous images which persists for a time after the initiating normal linear lightning. The luminous images are of the same apparent size as the diameter of a linear lightning channel and appear to be nearly spherically shaped. Each image is separated from the next by an unilluminated region. The dark gap may have dimensions of a few diameters of the illuminated segments

The bad lightning form is thought to evolve from an abnormal lightning channel between two clouds. The normal lightning discharge channel apparently decays into the series of disconnected luminous fragments. The complete bead lightning form, with its numerous segments, apparently occurs simultaneously and is not an image formed by a single moving illuminated object which has a periodic variation in brightness. It appears to observers to be a persistent illumination of a path followed by normal linear lightning and remains a relatively long time after the lightning flash has ceased. Lifetimes of such bead lightning occurrences are reported to be of the order of 1-2 sec.

Several specific characteristics distinguish bead lightning from normal lightning and ball lightning. A quasiwave structure formed by the discontinuous line image is often reported. It appears as several cycles of a sinusoidal line. The quasiwave structure is not always present, while the dotted appearance is the commonly reported characteristic of bead lightning." [1], *p.11/12.*

On a bead lightning photographed by two film cameras:

"Four successive lightning strokes were recorded. All strokes apparently followed the same path and each secondary discharge occurred before the previous discharge had completely dissipated. The beadlike appearance of the path appeared as each discharge faded. The bead structure was recorded by both cameras. The beadlike structure remained for about 0.3 sec." [1], *p.22.*

On another case of a photographed bead lightning:

"The stroke extended over 1 km in length and the bead persistence was 75 to 300 msec. The bead size was estimated at from 50 cm to several meters in diameter." [1], *p.25.*

Bead lightnings are very seldom observed:

"... only 22 occurrences in over 20 years of observation and 500.000 camera-hours of exposure time." [1], *p.25.*

On bead lightning origins:

"Bead lightning is thought to result from a lightning channel that develops a periodic longitudinal intensity oscillation due to some perturbing influence. The channel decays to become a series of almost uniformly separated areas with a persistent lumninosity ... . The luminous segments and the dark regions of the beaded channel are fairly regular ..." [1], *p.28/29.*

On explanations:

"The source of bead lightning is still an object of speculation." [1], *p.31.*

In Figure 6 a sketch is drawn of how a bead lightning looks:

Figure 6: Appearance of a Bead Lightning.

Same <u>data</u> of observations are:

<u>Bead persistence</u>:	75 to 300 msec.
<u>Bead size</u>:	50 cm to several metres.
<u>Bead intervals</u>:	2m to 50m?

3.2 What is Bead Lightning?

It is quite natural to assume that a Bead Lightning consists of a number of Ball Lightning equally spaced along the path of an ordinary lightning stroke. This idea has been repeated again and again in the literature. If one accepts the model of Section 2.4 for a ball lightning then most properties of a bead lightning are easy to explain:

They are mostly reported in higher altitudes where the existence of chemical fuel like swamp gas cannot be expected. Therefore they have **no light of their own** and are **illuminated only by**

(i) the luminosity of the initial plasma which cools off in less than one second and/or

(ii) the luminosity of one or more return strokes.

Also the size is expected to be in the metre range if these balls get their energy and air mass from an original jet of air of approximately 20 cm in diameter and if they are spaced at, say, 15 m intervals.

Hence, the only remaining question is, why these balls are equally spaced? This leads to the

MAIN PROBLEM FOR BEAD LIGTHNING

IS THERE A MECHANISM WHICH PRODUCES STANDING WAVES IN A LIGHTNING CHANNEL, FROM WHICH THE "PEARLS" OF BEAD LIGHTNING COULD EVOLVE ?

This can be explained by their development from a hydrodynamic instability which will be described in what follows.

3.3 The Birth of Bead Lightning

It has been speculated already earlier that the origin of a bead lightning is due to a certain instability phenomenon (see Section 3.1). There was, however, no hint that this could be a hydrodynamical instability.

In order to understand this problem better let's for the moment turn to a two-dimensional boundary layer of a viscous fluid at a wall. It was first proven vigorously by Tollmien that such a layer is in general instable with respect to two-dimensional waves, the so-called Tollmien waves. A very famous photograph of an experiment conducted by the late F.X. Wortmann is copied in Figure 7 (the original is a gift of F.X. Wortmann to the author). It shows the (water) flow (coming from the left) on a straight wall. By coloring the water at different heights above the wall the flow pattern can be observed by so called strokelines. Two Tollmien waves are clearly visible. Then the flow develops a rotational backward motion.

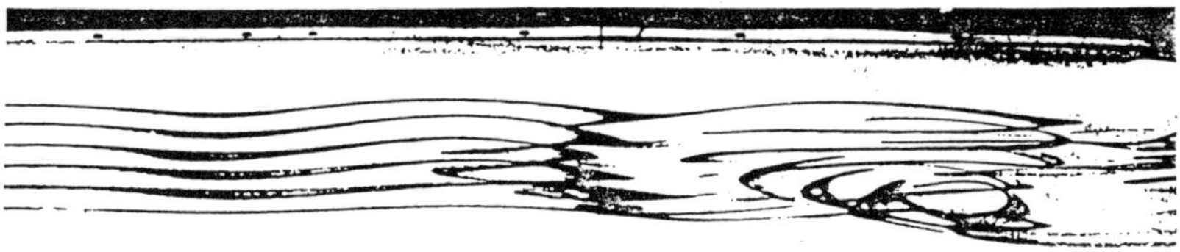

Figure 7: Development of Tollmien waves due to hydrodynamic instability and backflow. Main flow from the left. Copy taken from a photograph of the late F.X. Wortmann.

To show this clearly a copy of the second stroke line from below has been made. A second copy has then been placed upside down under it, thus producing a reflection of the wave pattern at the wall acting as a mirror. To the left the original initial velocity profile(s) can be seen. To the right (downstream) a pair of vortices have been formed as a secondary instability from the first order Tollmien wave instability.

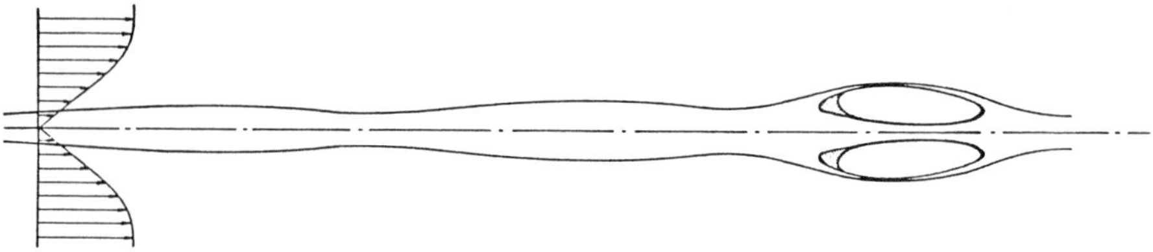

Figure 8: Second strokeline from below taken from Figure 7 and duplicated mirror-like. To the left: initial velocity profile(s).

This constructed picture can be taken as a hint for the development of a three-dimensional rotational symmetric jet. Kindly note that the pattern in Figures 7 and 8 is time independent. The development of Tollmien-like waves in a three-dimensional jet, originating from a lightning stroke, would in contrast to this be time dependent. A possible streamline (strokeline) pattern of such an instability is sketched in Figure 9. This is thought of as a momentary picture which can give a rough idea how the pearls of bead lightning may develop.

If there are one or more successive strokes the pearls would be strengthened by such an event.

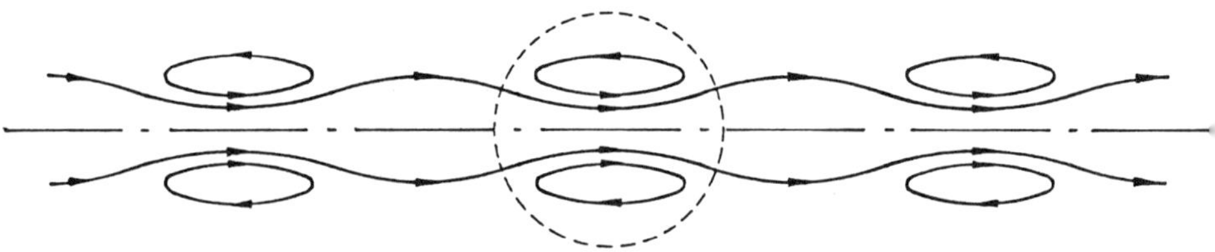

Figure 9: Sketch of the possible development of Bead Lightning due to hydrodynamic instability of the original jet of plasma and air after an initial lightning stroke.

3.4 Connection Between Bead Lightning and Ball Lightning

If one takes the phenomenon of bead lightning for granted then a quite "natural" explanation for ball lightning would be:

A Ball Lightning is one pearl from a Bead Lightning.

At first sight, all evidence seems to speak against that explanation. Singer [8], p.25 writes:

> *"The conversion of the bead form to typical ball lightning, for example, by the disappearance of all but one bead which then moves in a relatively complex path, has not been reported. It appears that despite the resemblance between the two forms, the occurrence of one excludes the other although the number of bead lightning observations on which this conclusion is based is small. The appearance of a single ball a few seconds after the disappearance of a bead lightning and in the same was reported in a unique case.*
>
> *The observer at first noted cloud-to-cloud strokes only. occuring at great frequency, 200 or more strokes per minute.A cloud-to-ground flash was then seen with several successive strokes in the same channel. The final stroke appeared especially intense, and it decayed to a chain of beads in the same channel. A few seconds after the beads had disappeared another stroke followed the same path, and shortly after, a glowing ball formed at the upper end of the channel under the cloud layer and floated slowly to earth. Following another strong stroke a bright mass of light appeared, but its descent to earth was too rapid to allow its identification as another ball. The slow motion of a luminous sphere along the channel of a lightning flash immediately after the disappearance of the continuous discharge was observed a few times in another storm with frequent lightning."*

But if one accepts the model for ball lightning given in the previous paragraph 2, and that of bead lightning given above there is no discrepancy: Bead lightning is normally observed only at larger distances from the ground. One would not expect to find very often inflammable gases, e.g., swamp gas, at higher altitudes. Therefore, the only "natural" light source for a bead lightning is that of the plasma formed in the preceding lightning stroke (and, sometimes, that of successive lightning strokes). As soon as the plasma gets too cool to radiate visible light anymore, the pearls of the bead lightning become invisible. The real reason for the beads, the hydrodynamic Hill vortices, still exist afterwards for a much longer time. But they are no longer illuminated and remain, therefore, hidden in the dark.

The case reported in Singer's book could very well be one of the very seldom encountered cases where an invisible Hill vortex has been fed chemical fuel by some mechanism (e.g., H_2 formed by hydrolysis of water droplets ?) and then became visible by a burning process as a ball lightning.

4. Final Remarks

The model for the nature of ball and bead lightning as given in this paper can in short be described as follows:

A ball lightning is primarily a sphere of rotating air, gas or plasma which is kept together by hydrodynamical forces as in a Hill vortex or in a smoke ring. Its luminosity and heat are produced by a chemical burning process of gases trapped inside the sphere. It is initiated by an ordinary lightning stroke which delivers its momentum, kinetic energy, chemical fuel and electrical charge.

A bead lightning consists of a string of ball lightning, normally without chemical energy. The regularity of its periodic structure comes from the hydrodynamic instability of the jet of air produced by its preceding lightning stroke.

Both phenomena are probably very common but are rarely observed since the conditions of "coloring" them are very seldom encountered.

This model is not entirely new. It consists of a combination of 3 well known, different mechanisms. Each of them has been - separately - discussed widely in the literature. What is new perhaps, is only the change of view that ball lightning and bead lightning are not regarded primarily as an electrical phenomenon or as one of chemical nature but that the primary causes for the observed structures are purely hydrodynamic forces. From this viewpoint the explanation of the periodic structure of bead lightning comes quite "naturally" from hydrodynamic instability.

Many **Mathematical_Questions** arise from this model. Among them are:

(i) What is the **lifetime** of Ball Lightning ? Using the Hill vortex as a model and boundary layer theory one gets an explicite formula and an estimate of up to several minutes for "usual" conditions. See the next paper of the author in this volume.

(ii) Describe the **conversion** of a jet hitting the ground into a ball-like structure.

(iii) Give a full description of the **stability** of an infinitely long time dependent, slowly dissipating jet. What is the critical Reynolds number and the wavelength for the equivalent of Tollmien waves? With the usual model and the observed data one gets a pearl size of up to one metre and a wavelength of up to 15 m. See the **Appendix** to this paper.

(iv) **What** is the **transition** from such waves to a fully developed bead structure?

(v) What happens by **repeating** the initial jet several times in analogy to successive lightning strokes?

There are also, of course, many physical and chemical questions which are still open and should be answered by both theoretical and experimental approaches. The final **goal** is :

> *BALL LIGHTNING AND BEAD LIGHTNING WILL NOT BE UNDERSTOOD COMPLETELY, UNLESS IT IS POSSIBLE TO CREATE THEM ARTIFICIALLY.*

5. Acknowledgements

This paper has been written while the author worked as a visiting fellow at the Centre for Mathematical Analysis of the Australian National University in Canberra/Australia. I wish to thank the Director of this Centre, Professor N.S. Trudinger for inviting me and for giving me the opportunity to do some research there. My thanks also go to my good friend, Professor R.S. Anderssen at CSIRO in Canberra for encouraging me to write down these ideas, I had been working at them for over four decades. I had much help from many people and got many tips from them for which I also wish to thank, especially to Professor L.E. Fraenkel for fruitful discussions and for his friendly advice.

APPENDIX

THE INSTABILITY OF A LIGHTNING CHANNEL

The lightning channel will be treated here as a jet of air which dissipates slowly, see Figure 10.

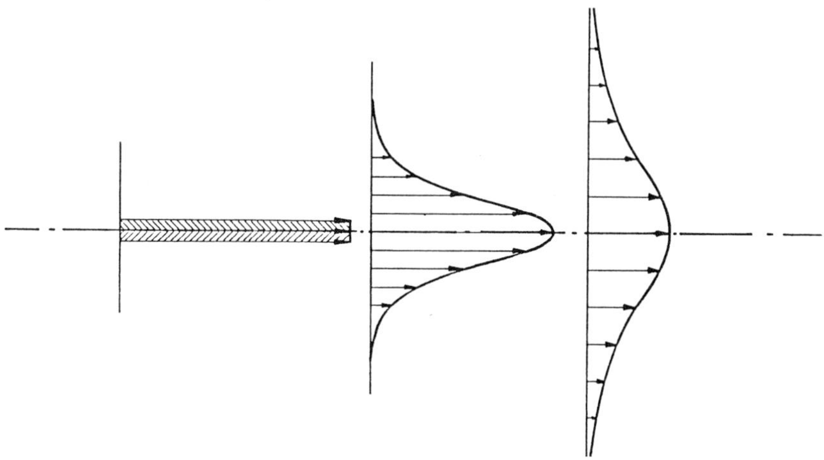

Figure 10: Lightning channel idealized as a jet of air. The initially approximately constant air speed in the channel will dissipate and, thus, create a bell shaped velocity distribution. Both jet flow direction and time development from the left to the right.

For simplicity first the **two-dimensional case** will be treated. Let x be the direction of the jet flow in the x,y coordinate system and denote by u,U and v the velocity components in the x- and y-direction, see Figure 11. Assume a two-dimensional flow of an incompressible medium with the kinematic viscosity ν.

An explicit, time dependent **solution** of the Navier-Stokes equation, which models the decay of an infinite jet, has the velocity component U in x-direction given by

$$U(t,y) := \frac{W}{\sqrt{\nu t}} e^{-\frac{y^2}{4\nu t}}$$

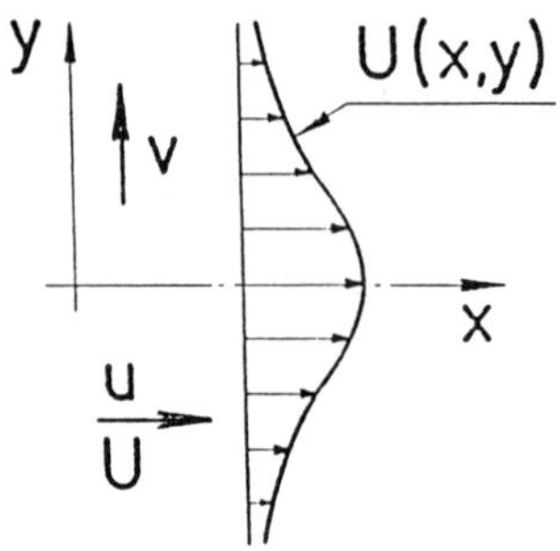

Figure 11: Coordinate system and velocity profile U(t,y) of the unperturbed flow.

Herein t is the time, $W = U(1/\nu, 0)$ is an appropriate constant; see Figure 11. Kindly note that U does not depend on x.

Now, the velocity components of the **perturbed flow** are written as $U(t,y) + u(t,x,y)$, $v(t,x,y)$ and a **stream function** $\psi(t,x,y)$ is introduced with the properties

$$u = \partial\psi/\partial y, \quad v := -\,\partial\psi/\partial x.$$

After omitting the pressure term and **linearization**, the Navier-Stokes equation then becomes the following **Perturbation equation**

$$\Delta\psi_t + U\Delta\psi_x - U_{yy}\psi_x = \nu\cdot\Delta^2\psi \qquad (A.1)$$

with the usual **boundary conditions** for ψ at $y=0$ and $y = \pm\infty$.

As in the case of **Tollmien waves** one expects to find a **critical Reynolds number**. Surprisingly enough, this is <u>not</u> the case, i.e. the solutions of equation(A.1)are **globally stable**. This can be shown, e.g., by expanding ψ in an appropriate power series where the coefficient functions can be expressed explicitely (!). It can also be shown by the following

Transformation:

Define: $\tau = \nu t$, $W^* = W/\nu$,

$$U^*(\tau,y) \mathrel{\hat{=}} U(t,y) = \frac{W^*\nu}{\sqrt{\nu t}}\, e^{-\frac{y^2}{4\nu t}},$$

then equation(A.1)becomes

$$\Delta\psi_\tau + U^*\Delta\psi_x - U^*_{yy}\psi_x = 1\cdot\Delta^2\psi,$$

i.e.: **the kinematic velocity ν can be choosen as 1 without loss of generality.**

But, as a supplement to this general result, the above equation(A.1)can been shown to be **locally instable**. This stems from the fact, that the function U changes **slowly** in time. For air flow at room conditions for example, the velocity U of the unperturbed flow is reduced to half of its value in approximately 10 seconds. Since the development of pearl lightning takes place in less than 1 second, the stability problem can be treated **quasi stationary**, i.e.: the main flow can be regarded as "frozen" with $U(t,y) \approx$ constant in time.

This immediately leads to the famous **Orr - Sommerfeld** equation and from this the existence of **Tollmien waves** can be proven and computed as usual.

Numerical computation*) gives a **critical Reynolds number** of the order of magnitude

$$Re_{crit} \approx 100 \quad .$$

The **wave length** is approximately*)

$$\lambda \approx 15\,\delta \quad ,$$

where δ is the **momentum loss thickness**.

---

*) To be published soon, after evaluating the values for the three-dimensional case, too.

Typical data for (3-dimensional) lightning channels are :

speed: 10 to 100 m/sec,

δ : 10 to 1000 mm,

hence the Reynolds number (for air at room temperature)

Re : 7 000 to 7 000 000

exceeds by far the critical value and the wave length

λ : 0.15 to 15 m

lies in the observed range of the distance between two "pearls" of bead lightning .

The **three-dimensional case** is essentially the same. The equations, however, become more complicated. The velocity distribution of the unperturbed flow is now

$$U(t,r) = \frac{W}{\nu t} e^{-\frac{r^2}{4\nu t}}$$

with the distance r from the centre of the jet. The **same conclusions** can be drawn as in the two-dimensional case. Also, the **critical values** have the same order of magnitude.

At the Conference the computations were not finished. The complete theory together with numerical results will be published elsewhere.

REFERENCES

[1] **Barry, James Dale**: Ball Lightning and Bead Lightning. Plenum Press 1980.

[2] **van Dyke, Milton**: An Album of Fluid Motion. Stanford, Parabolic Press 1982, p.44.

[3] **Fraenkel, L.E.** and **M.S. Berger**: A Global Theory of Steady Vortex Rings in an Ideal Fluid. Acta Mathematica 132 (1974), p.13-51.

[4] **Hassenstein, Werner**: Kugelblitz - direkt was Gemütliches. Constanze 7 (1954), No.5, p.5.

[5] **Lamb, Horace**: Hydrodynamics. Cambridge, University Press, reprinted 1962, p.245/6.

[6] **Norbury, J.**: A Family of Steady Vortex Rings. J. Fluid Mech. 57 (1973), p.417-431.

[7] **Schneider, Justus**: Medizinisches vom Gewitterflug am 6. August 1938. Thermik 3 (1950), p.19/20.

[8] **Singer, Stanley**: The Nature of Ball Lightning. Plenum Press 1971.

THE LIFETIME OF HILL'S VORTEX

Karl L.E. Nickel

Institut für Angewandte Mathematik
Universität Freiburg
Freiburg i.Br.
West Germany

ABSTRACT

Hill's vortex is an explicit solution of Euler's equation for a steady flow of an incompressible and inviscid fluid. It describes the flow around and in a ball in a parallel flow with constant velocity in infinity. It is similar to the flow pattern of a smoke ring. In a "real", i.e., viscous, fluid such a steady motion is impossible, hence, the Hill vortex will decay. By using a boundary layer approximation and a one-dimensional time dependent explicit Navier-Stokes solution the decay rate and the lifespan of a Hill vortex can be approximated. A lower estimate for this decay time T is

$$T = \frac{\pi}{225} \frac{d^2}{\nu} ;$$

where d is the ball diameter and ν the kinematic viscosity of the fluid.

AMS MOS: Primary 76C05 Secondary 35K15

This paper first appeared as :
CMA Research Report CMA-R12-86; March, 1986 at the Centre for Mathematical Analysis of the Australian National University in Canberra/Australia.

1. Introduction

Quite recently the author suggested Hill's vortex [1] as a primary model for the understanding of the nature of ball lightning [2]. In order to know something about the decay of ball lightning, therefore, first the lifetime of Hill's vortex has to be investigated. This is done in the following paper.

2. The Hill Vortex

In 1894 Hill introduced a flow field which by now is known as the "Hill Vortex":

Consider the steady motion of an inviscid and incompressible fluid in the three-dimensional x,y,z-space.

Define $r \geq 0$ by

$$r^2 = x^2 + y^2 + z^2$$

and $\varrho \geq 0$ by

$$\varrho^2 = y^2 + z^2 .$$

Let $a > 0$ and $W > 0$ be given real constants. Define the streamfunction $\psi(x,y,z) \mathrel{\hat{=}} \psi(\varrho,r)$ by

$$\psi(\varrho,r) = \begin{cases} \frac{3}{4}W\varrho^2\left(1 - \frac{r^2}{a^2}\right) & \text{for } r \leq a \\ -\frac{1}{2}W\varrho^2\left(1 - \frac{a^3}{r^3}\right) & \text{for } r \geq a . \end{cases} \tag{1}$$

Then this streamfunction describes the fluid flow around a sphere of radius a with the fluid moving at infinity with the constant velocity $W > 0$ in the positive x-direction. Some of the streamlines of this flow in the x,y-plane for $z = 0$ are sketched in Figure 1, which is taken from the book of Lamb [1], page 246.

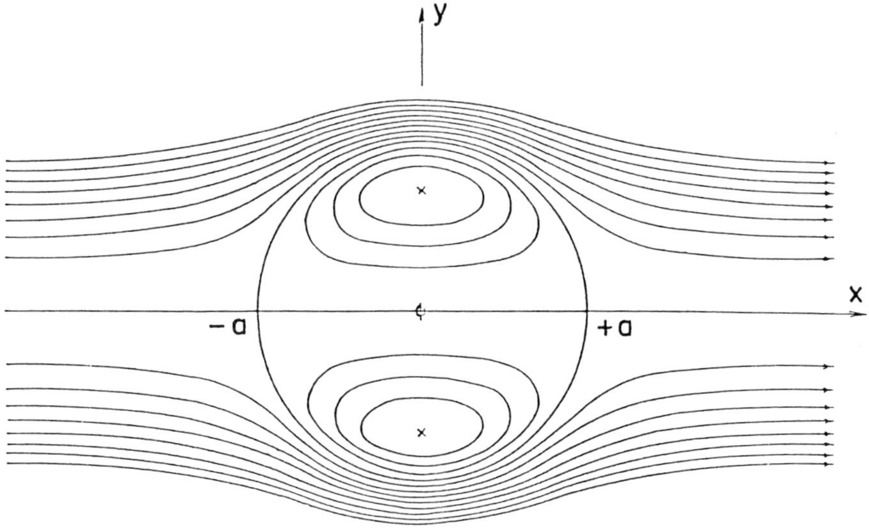

Figure 1: Fluid flow around a sphere. The streamlines in the x,y-plane are drawn for equidistant values of ψ. From Lamb [1], p.246.

The velocity components $U = U(x,y,z)$ in x-direction and V in ϱ-direction can be evaluated from ψ by the well known expressions

$$U = -\frac{1}{\varrho}\frac{\partial\psi}{\partial y}\,, \quad V = \frac{1}{\varrho}\frac{\partial\psi}{\partial x}\,.$$

At the line $x = z = 0$, hence $r = \varrho = y$, one finds easily

$$U(0,y,0) = \begin{cases} -\frac{3}{2}W\left(1 - 2\,\frac{y^2}{a^2}\right) & \text{for } 0 \le y^2 \le a^2 \\ \frac{1}{2}W\left(2 + \frac{a^3}{y^3}\right) & \text{for } y^2 \ge a^2\,. \end{cases}$$

At $y = a$, i.e., at the ball boundary, this equation gives the value

$$U(0,a,0) = \frac{3}{2} W.$$

At this line the x-component U of the velocity therefore has a value which is by 50% higher than that at infinity. This "excess velocity" is due to the displacement effect of the sphere.

The velocity profile $U(0,y,0)$ has been sketched in Figure 2 for $y \geq 0$.

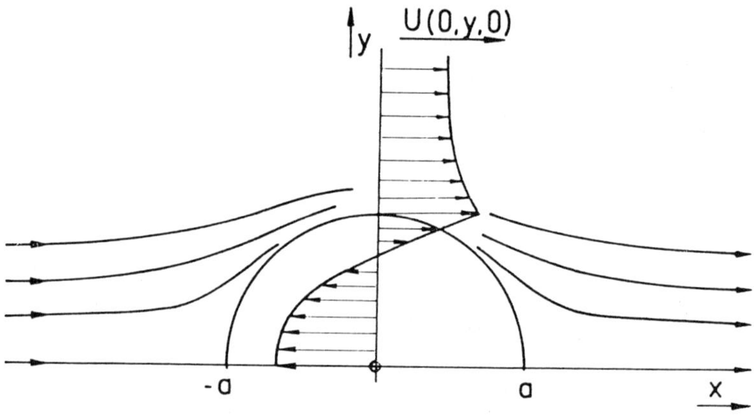

Figure 2: The velocity component $U(0,y,0)$ in x-direction of the Hill vortex.

From this Figure 2 one sees clearly the well known property, that the velocity component U is continuous at the boundary $r = a$ of the sphere but that the gradient $\partial U/\partial y$ is <u>not</u>! Hence there is a jump in the shear stress for $r = a$. Therefore the fluid flow generated by equation (1) can not be a solution of the unsteady Navier-Stokes equation of a viscous fluid. One expects, therefore, that (1) describes the flow only at the initial time and that thereafter a time dependent flow develops as a solution of the unsteady Navier-Stokes equation.

By analogy to the behaviour of a smoke ring one expects that the Hill vortex decays in time until it is completely dissipated due to viscositiy. How fast is this decay rate? To the knowledge of the author this question has not yet been answered. In what follows a rough approximation will be made using some ideas from boundary layer theory and an explicit solution of the time dependent Navier-Stokes equation.

3. A Boundary Layer Approximation of the unsteady Hill Vortex

In Figure 3 the velocity equation component $U(0,y,0)$ of Hill's vortex at the equator is sketched again for $y \geq 0$.

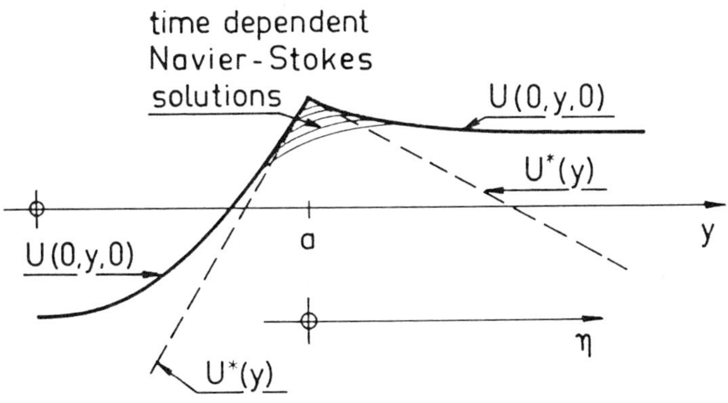

Figure 3: The velocity component $U(0,y,0)$ together with a linear approximation $U^*(y)$ at $y = a$ and with some solutions of the Navier-Stokes equation for different times.

Also incorporated in this picture are some solutions of the unsteady Navier-Stokes equation. For increasing time they have the tendency to "smooth out" the corner of $U(0,y,0)$ at the surface $y = a$ of the sphere.

If one uses the new coordinate

$$\eta = y - a$$

then the linear approximation $U^*(y)$ of $U(0,y,0)$ at the point $y = a$ becomes

$$U^*(y) \mathrel{\hat{=}} U^*(\eta) = \begin{cases} \frac{3}{2}W(1 - \frac{\eta}{a}) & \text{for } \eta \geq 0 \\ \frac{3}{2}W(1 + 4\frac{\eta}{a}) & \text{for } \eta \leq 0. \end{cases} \tag{2}$$

It is the goal of this paper to construct the development of the flow of a viscous fluid starting with Hill's vortex as an initial condition. While Hill's vortex represents a solution in closed form of Euler's equation no such solution is known for the time dependent Navier-Stokes equation. In what follows, therefore, some approximations have to be made. They consist of 3 simplifications:

i) The fluid flow is divided into a "main part", which is regarded as given and into a secondary part which is to be computed. This is the standard boundary layer approach. In what follows the steady flow produced by Hill's vortex will be used as the basic flow. One may assume that for small time intervals fairly good results can be expected.

ii) It would be rather difficult to compute a boundary layer approximation for the flow near the whole surface $r = a$ of the ball. Therefore, only the equation $x = 0$, $r = a$ is considered. This region is regarded as the most "endangered" part of the flow and also as that with the most "typical" flow behaviour.

Hence, instead of computing the flow at the whole two dimensional ball surface $r = 0$ only a one dimensional approximation at $x = 0$ will be considered. This change will be the most incisive simplification made in what follows.

iii) It seems that there is not a solution of the boundary layer equation to the primary flow $U(0,y,0)$ which can be expressed in closed form. Hence for simplificity the velocity profile $U(0,y,0)$ will be replaced by the linear approximation $U^*(y)$.

It is rather surprising and comes as a bonus that there is even a closed form solution of the full Navier-Stokes equation to this basic flow.

The difference between U and U^* is small close to $r = a$ where it really counts for small time intervals. Therefore this third simplification will not affect much the solution for small time intervals.

All these three simplifications will produce a decay rate which is too fast. One may conclude herefrom that the final results will be too pessimistic. Hence the final formula for the lifespan of Hill's vortex should give too small values.

4. A Similar Solution of the Navier-Stokes Equation

The system of Navier-Stokes equations for the time dependent, two-dimensional flow of an incompressible viscous fluid in the x,η-plane with the time t is

$$u_t + uu_x + vu_\eta + p_x - \nu(u_{xx} + u_{\eta\eta}) = 0 \qquad (3)$$

$$v_t + uv_x + vv_\eta + p_\eta - \nu(v_{xx} + v_{\eta\eta}) = 0 \qquad (4)$$

$$u_x + v_\eta = 0. \qquad (5)$$

Here $u(t,x,\eta)$, $v(t,x,\eta)$ are the two velocity components in x,y-direction; p is the pressure and ν the kinematic viscosity. Without loss of generality the fluid density is taken as 1. As always, letters as indices denote partial derivation. Because of the model presented in Section 3 a solution to (3), (4), (5) is wanted for $t \geq 0$ in $-\infty < x,\eta < \infty$ which satisfies the initial condition

$$\left\{ \begin{array}{l} u(0,x,\eta) = U^*(\eta) \\ v(0,x,\eta) = 0 \end{array} \right\} \quad \text{for } -\infty < x,\eta < \infty. \qquad (6)$$

There are some additional conditions at infinity and for $p(t,x,\eta)$, which will be omitted here. The variable x does not appear explicitely in the initial condition (6), in the flow region nor in the equations (3) to (5). Hence one can expect that there is a solution u,v,p which is independent from x. By the continuity equation (5) and the initial condition (6) this leads to $v(t,x,\eta) \equiv 0$. Hence the equations (4) and (5) are both satisfied for $p_\eta = 0$ which leads to

$p(t,x,\eta) \equiv 0$. Therefore the system (3), (4), (5) reduces to the well known heat equation

$$u_t - \nu\, u_{\eta\eta} = 0. \tag{7}$$

By defining

$$\varphi(t,\eta) = U^*(\eta) - u(t,x,\eta) \tag{8}$$

and observing that U^* is a solution of (7) one gets finally the equation

$$\varphi_t - \nu\, \varphi_{\eta\eta} = 0 \quad \text{for} \quad 0 < t,\ -\infty < \eta < \infty \tag{9}$$

for $\varphi(t,\eta)$ with the initial condition

$$\varphi(0,\eta) = 0 \quad \text{for} \quad -\infty < \eta < \infty, \tag{10}$$

the boundary condition

$$\varphi(t,\pm\infty) = 0 \quad \text{for} \quad 0 < t \tag{11}$$

and the jump condition at $\eta = 0$

$$\varphi_\eta(t,+0) - \varphi_\eta(t,-0) = -\frac{15}{2}\,\frac{W}{a} \quad \text{for} \quad t > 0. \tag{12}$$

The last condition comes from the fact that

$$U^{*\prime}(\eta) - U^{*\prime}(-\eta) = -\frac{15}{2}\,\frac{W}{a} \quad \text{for all} \quad \eta \geq 0.$$

By using the known theory for the heat equation the following uniquely determined solution to (9), (10), (11), (12) can be found:

$$\varphi(t,\eta) = \frac{15}{2}\,\frac{W}{a}\,\sqrt{\nu t} \int_{\frac{|\eta|}{2\sqrt{\nu t}}}^{\infty} [1-\mathrm{erf}(s)]\, ds \tag{13}$$

with the well known error function

$$\operatorname{erf}(s) = \frac{2}{\sqrt{\pi}} \int_0^s e^{-t^2} \, dt.$$

The solution (13) is obviously a "similar solution", i.e., it has the form

$$\varphi(t,\eta) = g(t)h(\eta/g(t))$$

with the obvious choice $g(t) = \sqrt{2\nu t}$ and an appropriate choice for $h(s)$. It seems, that this explicit solution of the Navier-Stokes equation has not yet been known (?).

5. The Lifetime of Hill's Vortex

The actual decay of Hill's vortex is unknown up to now. It is approximated here by the model presented in the two previous sections. One may expect that this model describes the flow behaviour quite good at $x = 0$, $r = 0$ and for small time intervals. It can only be speculated on how good it describes the real phenomenon at the last stages of the existence of Hill's vortex. One has, therefore, to make a guess at what time Hill's vortex is terminated and cedes to exist. Here the following assumption will be made:

> *The Lifetime T is defined as that time, after which the "excess velocity" $W/2$ at the equator of Hill's vortex is eaten up by viscosity.*

By inserting this definition in equations (8), (2) and (13) one gets the equations

$$\frac{W}{2} = u(t,0) = U^*(0) - \varphi(t,0)$$

$$= \frac{3}{2}W - \frac{15}{2}\frac{W}{a}\sqrt{\nu T}\int_0^\infty [1-\mathrm{erf}(s)]ds$$

or

$$T = \frac{4}{225}\frac{a^2}{\nu(\int_0^\infty [1-\mathrm{erf}(s)]ds)^2} .$$

An elementary analysis produces the constant

$$\int_0^\infty [1-\mathrm{erf}(s)]ds = \frac{1}{\sqrt{\pi}} .$$

Inserting this and replacing 2a by the diameter d gives the final formula

$$T = \frac{\pi}{225}\frac{d^2}{\nu} \qquad \text{with} \quad d = 2a. \tag{14}$$

As explained before this value is most probably too small and a lower bound for the actual life expectancy of Hill's vortex.

Define the (dimensionless) Reynolds number Re by

$$Re = \frac{dW}{\nu} .$$

By using dimension analysis one sees immediately that the life-time of Hill's vortex has to have the following structure

$$T = c(Re)\frac{d^2}{\nu}$$

with a suitable function c(Re). Hence the above approximation theory and formula (14) yields the extremely simple result

$$c(Re) = \frac{\pi}{225} = \text{const.}$$

Experiments seem to support this result that the lifetimne T of a physical vortex ring does not depend upon the velocity W, i.e. that the function c(Re) really is constant.

Discussion:

1) It may come as a big surprise that the velocity W at infinity does not affect this decay time T . Hence a Hill vortex may have any speed W, it will nevertheless always "die at the same age".

2) The dependence of T on a and ν is as expected:

   i) Large Hill vortices live longer.

   ii) Small values of the kinematic viscositiy ν also give a high life expectancy; i.e., a Hill vortex in hot air will decay faster than in cold air, and such a vortex in water will live approximately 13 times (!) longer than one of the same size in air.

3) Two short tables of the values of the kinematic viscosity ν (in the c-g-s-system) show these effects more clearly.

Table I: Kinematic viscosity ν in cm^2/sec for <u>air</u> at atmospheric pressure for different temperatures. Also given is the ratio T/T_0 of the lifespan T of a Hill vortex to that T_0 at 15°C temperature.

Temperature in °C	-100°	0°	100°	200°	300°
ν	0.057	0.132	0.230	0.346	0.478
T/T_0	2.56	1.10	0.63	0.42	0.30

Table II: Kinematic viscosity ν in cm^2/sec at atmospheric pressure and 15°C temperature and ratio T/T_0 of the lifespan T of a Hill vortex compared to that T_0 in air.

Fluid	Air	Water	Paraffine	Glycerine
ν	0.1458	0.0114	0.25	10
T/T_0	1	12.8	0.58	0.015

4) For air at atmospheric pressure and at room temperature formula (14) becomes simpler. Using the value $\nu = 0.1458$ cm^2/sec and approximating the number 10.442... by 10 one gets:

The lifetime of Hill's vortex in air at atmospheric pressure and at room temperature is approximately

$$T \approx \frac{d^2}{10} \text{ in sec}$$

with the diameter d in cm.

The evaluation of this formula for some diameters d is given in Table III.

Table III: Lifespan of Hill's vortex in air

Diameter d in cm	3.16	10	20	30	50	100
Lifespan T	1 sec	10 sec	40 sec	1'30"	4'10"	16'40"

The observed lifetime of smoke rings with a diameter d of 8 to 10 cm's is normally about half a minute. If one considers their (probably) different mechanism and the fact that formula (14) gives only a rough lower bound this value is not too far away from the 10 seconds predicted in Table III.

It would be interesting to use the method of differential inequalities to get guaranteed lower and upper bounds for T.

Acknowledgements

This paper has been written while the author was a visiting fellow at the Centre for Mathematical Analysis of the Australian National University in Canberra/Australia. I wish to thank the Director of this Centre, Professor N.S. Trudinger, for inviting me and for giving me the opportunity to do there some research.

The stimulus to this paper came from Professor L.E. Fraenkel who gave a Seminar at the Centre on vortex rings. I wish to thank him very much for several fruitful discussions, for his friendly remarks and for his advice.

References

[1] Lamb, Horace: Hydrodynamics. Cambridge, University Press. Reprinted 1962, p.245/6.

[2] Nickel, Karl: A Model for Ball Lightning and Bead Lightning. CMA Research Report CMA-R11-86; February, 1986.

Electrical and radiative properties of ball lightning.

Smirnov B.M.

Uspekhi,Leninskij prosp. I5, Moscow,II707I,USSR.

I.Introduction.

Ball lightning is an interesting natural phenomenon which attracts the attention of scientists during long time.In a result of investigations fulfilled (I - 4) now we have the reliable representation about observational properties of ball lightning. Table I the parameters of the oservational model of ball lightning,i.e. the average parameters of ball lightning.These data give the representation about observational ball lightning and composite the basis for the following analysis of its nature. In such analysis (5-8) it is nessary to take into consideration that ball lightning is many-sided phenomenon and each aspect of this phenomenon is connected with its range of physical processes and phenomena.The contemporary investigations of the physical nature of ball lightning con - sist of a study and modelling of its separated aspects.The whole complex of such investi gations gives some phenomenological model of ball lightning.The detailed study of sepa - rated aspects of the ball lightning problem makes more profound our representation about this phenomenon and in the end will allow to model ball lightning as a whole. In this paper which is the development of (8) we stop on two aspects of phenomenon considering the electrical and radiative properties of ball lightning.

Table I .

Parameters of middle ball lightning according to observational data (9).These data use observations in USSR and USA.In Japan (I0)usually the origin of ball lightning does not connect with the thunder-storm weather.

	Parameters	Value
I	Spherical form probability	89 ±I %
2	Diameter (cm)	28 ±4
3	Lifetime (sec)	$I0^{0.95 \pm 0.25}$
4	Energy (kJ)	$I0^{I.3 \pm 0.2}$
5	Light stream (lumen)	$I0^{3.I \pm 0.2}$
6	Light efficiency (lum/Wt)	$0.7 \cdot I0^{\pm 0.65}$
7	Colour	White(24 ± 2%),yellow(24 ± 2%),red(I8 ± 2%),orange(I4 ± 2%),blue and violet (I2 ± I%) and others

8	Correlation with atmospheric electrical phenomena	70 ± I0% of ball lightnings are observed in the thunder-storm weather
9	Decay	50 ± 20% by explosion,other cases by slow extinction or breaking up into pieces
I0	Probability of the observation during the lifetime of a person	$I0^{-4 \pm I}$

One of the speciality of ball lightning is the chemical way of the internal energy storage.This conclusion follows ([II]) from small times of the storage for other energy types and from the large value of the special capacity of the chemical energy comparing with other types.Due to the complication of the process of the chemical energy transformation to thermal and radiative ones now we can not suggest concrete chemical compositions of ball lightning;

The ball lightning problem of principle respects to its structure.Leningrad physists ([I2]) supposed that ball lightning has the thread-like structure.It is the only way which allows to explain the conservation of its form during the evolution and does not contradict with other aspects of this phenomenon.Authors gone to this hypothesis from analysis of their experiments ([I3]) on an electrical explosion of metallic wires. Such structures were formated at definite regimes of the process.Formating clusters attached to camera walls and could be found in such state some days.But more detailed analysis ([I4]) has showed that by an electrical explosion of materials a fractal structure is formated. Hence one can conclude that the framework of ball lightning has the fractal structure. The fractal structure of the framework is more informative comparing with the thread-like one.The thread-like structure is a rough approximation of a fractal one.

It is essential that the ball lightning structure considered explain simultaneuosly a character of the ball lightning origin.Indeed,the ball lightning framework is formated as a result of sticking together of hard small particles which formate the friable structure by joining with each other.A study of kinetics of this process ([8;I5]) allows to estimate a size of these particles which is of order I - I0 nm.

The usage of the fractal structure of ball lightning together with other considerations allows to obtain the additional information about ball lightning parameters.Such parameters,which does not connect with its electrical and radiative properties considered further,are included in Table II ([8,I5]).

Table II.

The estimated parameters of ball lightning [8,15].

	Parametr	Value	The basis of estimation obtaining
I	The weight of a chemical active substance of ball lightning	2 - 10 g	By usage of the energy of middle ball lightning (20 kJ) and the specific energy storage of the active substance (2 - 10 kJ/g)
2	The ratio of ball light-weight to the air weight inside it.	$1\cdot 10^{\pm 0.8}$	The condition of a floating of ball lightning
3	A middle air heating inside a ball lightning framework	$60K\cdot 10^{\pm 0.6}$	The calculation of heat transfer due to a convective flow of an air through the framework.
4	A size of a particles in the ball lightning framework.	I - 10 nm	The small time of the framework formation
5	The temperature of the radiative zones.	$<$ 4 000K	A calculation of the temperature for products of gas mixture combustion.

2.Electrical properties of ball lightning.

According to observational data ball lightning carries an electrical charge and has electrical properties.Indeed, in 16 - 20 % of observational cases [1,2,4] ball lightning moves along of conductors.In most cases,when was fixated its origin,ball lightning arised from electrical devices or metallic objects,sockets,radioreceiversand televisor telephons,radiators etc.Ball lightning causes a destroy of electrical circuts and systems,its action on the man is likewise to the action of an electrical discharge.Thus the presence of electrical properties of ball lightning is a fact.

The electrical charge of ball lightning creates its surface tension,which is equal

$$\alpha = \frac{q^2}{4\pi R^3} = \frac{E^2 R}{4\pi} \qquad (I)$$

Here R is a ball lightning radius,q is its charge, E is the electrical field strength near its surface.Due to the surface tension the electrical charge of ball lightning ensures the stability of its construction ([12])by averting of its structure destroying.

Make an estimation of electrical parameters of middle ball lightning by assuming that the weight of the framework of ball lightning is of order of an electrical force of its interaction with metallic objects.Corresponded parameters for middle ball lightning are represented in Table III.It is interesting that in the range of surface tensions values for middle ball lightning (0.03 - I J/m^2) is found the water surface tension at room temperatures (0.07 J/m^2).Note that the electrical energy of middle ball lightning is smaller its total energy in six orders (see Tables I and III).The electrical energy of ball lightning is approximatelyfife order lower than one,which can cause the man death under an action of an electric current ([16]).Thus an electrical charge of ball lightning is not dangerous for a man.But an electrical charge is essential for the existing of ball lightning ([12]) because it creates the stability of ball lightning.

Table III.

The electrical parameters of middle ball lightning.

	Parameter	Value	Accuracy
I	Charge,C	$8\cdot 10^{-7}$	$10^{\pm 0.5}$
2	Charge density, e/cm^3	$4\cdot 10^{8}$	$10^{\pm 0.3}$
3	Surface tension, J/m^2	0.2	$10^{\pm 0.8}$
4	Ratio of a charge to a mass, C/g	$5\cdot 10^{-8}$	$10^{\pm 0.5}$
5	Electric field strength near a surface,V/cm	$4\cdot 10^{3}$	$10^{\pm 0.6}$
6	Electrical potential, V	$5\cdot 10^{4}$	$10^{\pm 0.6}$
7	Electrical energy, J	0.04	$10^{\pm 0.9}$
8	Electrical pressure to a surface, Pa	I.2	$10^{\pm 0.6}$

Let us analyse processes of hard particles aggregation which give the formation of the ball lightning framework.These processes take place in an unipolar plasma and for this reason the framework of ball lightning obtains an electrical charge.Table IV contents times of processes of an unipolar plasma evolution.Choosed parameters of processes can correspond to a real situation.According to Table IV data the first stage of these processes is a charge transfer from a plasma to hard particles.The aggregation of hard particles leads to the formation of a charged cluster.It is essential that its charge does not influence on rates of an aggregation process.Further we shall make corresponded estimations.

Table IV.

Times of electrical processes and aggregations processes in an unipolar atmospheric plasma.The charge density is $I0^{9}$ e/cm^{3},the particles radius is equal 3 nm and the content of particle material in an air is I g/g.

	Process	Time, sec
I	Charging of particles	$2.5\cdot I0^{-3}$
2	Plasma decay as a result of the attachment of ions to particles	$3\cdot I0^{-I0}$
3	Recombination of positive and negative ions	$5\cdot I0^{-4}$
4	Separation of charges in a plasma under an action of an external electrical field I kV/cm	$I0^{-2}$
5	Aggregation of particles to a cluster	$I0^{-2}$
6	Discharging of the cluster framework in an atmospheric plasma	$I0^{3}$

The mechanism of the hard particles aggregation into a large cluster is a following ([II]). At the first stage of a process particles are joined into small clusters and further small clusters are joinned with each other into a large cluster ([I7,I8]). This process is finished if clusters occupy the total space or other mechanisms of the aggregation are included, for example, the approachment of clusters proceeds in a result of clusters falling under an action of their weight.

At the first stage of the process the charge of aggregating particles is inessential if the energy of the Coulomb interaction of clusters is smaller than their thermal energy T :

$$\frac{q^2}{R} \ll T \qquad (2)$$

where R is a size of a cluster. Give numerical estimations. The cluster mass is equal ([I9])

$$m = m_0 \left(\frac{R}{a}\right)^D \qquad (3)$$

where m_0 is the mass of individual particle, a is its radius, D is the fractal dimension of a cluster, which is equal to I.77 for the cluster - cluster regime of the aggregation (see [9]). For the density of a material of particles $\rho_0 = 2\ g/cm^3$ and the ratio of a cluster charge to its mass according to Table III data we have on the basis of formula (2) $R \ll 400$ mcm for a = 3 nm and $R \ll I0$ mcm for a = 0.I mcm. In the first case clusters of sizes of order of $R \sim I$ mcm occupy the total space, i.e. the framework formation is ceased earlier than a cluster charge becomes essential. In the second case for $R \gtrsim 4$ mcm it is essential other mechanism of the aggregation of clusters, which connects with clusters falling in the gravitational field ([20]). Then instead of (3) other condition is demanded :

$$\frac{q^2}{R^2} \ll m\, g \qquad (4)$$

i.e. the force of the Coulomb interaction of clusters is lesser than their weight. The condition (4) violates only for clusters sizes of order of a ball lightning size. Thus according to this analysis the electrical charge of clusters, which are joined in a framework of ball lightning, does not influence on the rates of aggregation processes.

Make this analysis of other way. Let us find particles sizes, for which the formation of the framework of ball lightning shall be over earlier than shall be fulfilled the condition $\frac{q^2}{R} = T$. In other words, the Coulomb interaction energy is smaller than thermal one for clusters which sizes of order of a correlation radius of the ball lightning framework (or maximum sizes of pores in the framework). From formula (3) we have for a correlation radius of the framework:

$$R_c = a \cdot \left(\frac{\rho_o}{\bar{\rho}}\right)^{\frac{1}{3-D}}$$

where ρ_o is the density of a framework material, $\bar{\rho}$ is the average density of this material in the volume. It allows to rescribe the condition $\frac{q^2(R_c)}{R_c} << T$ in the form:

$$\frac{q_o^2}{a} << T \left(\frac{\bar{\rho}}{\rho_o}\right)^{\frac{2D-1}{3-D}} \qquad (5)$$

where q_o is the average charge of an individual particle and $q_o \sim a^3$.

For room temperatures and $\rho_o = 2 - 3\ g/cm^3$ we have on the basis of Tale III data for $D = 1.77$ $a < 50nm \cdot 10^{\pm 0.6}$, for $D = 2$ $a < 12\ nm \cdot 10^{\pm 0.8}$, for $D = 2.1$ $a < 5nm \cdot 10^{\pm 0.9}$. Note that $D = 1.77$ respects to a cluster - cluster aggregation regime and to the Brownian motion of particles and clusters. In this case for particles sizes; which respects to ball lightning parameters (see Table III), one can assert that an electrical charge of particles and clusters does not influence on the formation of the ball lightning framework.

The analysis of rates of electrical processes (see Table IV) shows that times of charging of a particle and a cluster in a considered plasma is smaller than the time of the formation of the ball lightning framework. The equilibrium charge of a particle or a cluster in a quasineutral plasma is proportional to their radius, if this radius exceeds the free path length λ molecules in a plasma (for an atmospheric air λ = 0.06 mcm) The equilibrium charge of a perticle or a cluster in a quasineutral plasma is determined by the Fuks formula [21]:

$$q = \frac{RT}{e} \ln\left(\frac{\mathcal{D}_+}{\mathcal{D}_-}\right), \qquad (6)$$

where $\mathcal{D}_+, \mathcal{D}_-$ are the diffusion coefficients of positive and negative ions in a plasma T is the gas temperature. This formula is valid for the conditions $q \gg e$ and $R \gg \lambda$ and can be used for an analysis of the real situation.

Ions sorts in a real air and their diffusion coefficients depend on the air humidity and other its parameters. Using data [22] $\mathcal{D}_+ = 0.029\ cm^2/sec$, $\mathcal{D}_- = 0.043\ cm^2/sec$ and data [23] $\mathcal{D}_+ = 0.028\ cm^2/sec$, $\mathcal{D}_- = 0.036\ cm^2/sec$ we obtain for an atmospheric air on the basis of formula (6) :

$$\frac{q}{R} = (-6 \pm 1) \cdot 10^4 \ e/cm \qquad (7)$$

From this follows that the equilibrium charge of the ball lightning framework is approximately $10^6 e$, but according to Table III data it is equal $3 \cdot 10^{11} e$. Therefore we conclude that large charge of the ball lightning framework can be conserved during the aggregation, if this process proceeds in unequilibrium plasma.

Remark that the number density of ions in an atmosphere is of order of $10^2 - 10^3 e/cm^3$, but according to Table III data the mean number density of a charge in the ball lightning framework is higher than this one some orders. It means that the framework formation takes place in a plasma which density is higher than atmospheric one. (This fact is accounted in consideration by choosing of initial conditions for Table IV). For an explanation of this situation one can composite the following scheme of the formation of ball lightning framework. Through the given region of atmosphere flows an electric current which is caused by an external electric field. It creates respectively high degree of the air ionization and causes the charges separation.In such plasma is ejected a vapour of a framework material.A vapour is condensed in particles which are joined to clusters.Since the process proceeds in an unipolar plasma,formating particles and clusters gain an electrical charge.

For an analysis of processes of the charged cluster formation let us consider a simple model of the creation of iniatial conditions for the growth of such cluster.Assume that an electrical current flows from an air through the surface of an hard material.This current evapourates a surface material and creates a plasma which contents hard particles. The condition of the electric current continuity on the interface gives :

$$E_I \sigma_I = E_2 \sigma_2, \qquad (8)$$

where E_I, E_2 are the electric field strengths in an air and in a solid, σ_I, σ_2 are their conductivities.From this relation it is follows that due to different conductivities of an atmospheric plasma and a solid material the electric field jump arises on the interface.It creates the electrical charge on the interface.

According to considering model the surface charge is found in a thin lauer and is ejected in an atmospheric plasma during a surface evaporationunder an action of an electric current.Then an unipolar plasma is formated.The plasma charge transfers fast to hard particles.Thus on the first stage of the cluster formation we have an air with charged hard particles and the total charge of particles differs from zero.

Let us make some estimations on the base of the above model.For simplicity we assume that the air conductivity is small comparing withthe material conductivity,i.e.the electric field jump on the interface coincides with the electric field strength E_I in an air Assume that the square of the interface exceeds essential the cross section of ball lightning,i.e.we are limited by the one-dimensional case.Then according to the Poisson equation we have:

$$E_I = 4\pi e \int N\,dx$$

where N the charge number density on the interface, x directs perpendicular to the interface.Introducing R - an unipolar plasma size after the material evaporation we obtain for the number density of an excessive charge in a plasma :

$$N = E_I/(4\pi e R)$$

Let us take from a plasma a ball of a radius R. The charge inside this ball is equal :

$$q = 4\pi R^3 N e/3 = E_I R^2/3$$

Let us assume that hard particles are joined in a cluster of the radius R. Then this cluster has a charge q and creates near its surface an electrical field of the strength

$$E = q/R^2 = E_I/3 \qquad (9)$$

Thus the electric field strength of charged cluster is of order of electrical field strength in an initial air through which flows an electric current.The totality of processes,which respect to the framework assembling,is such that the electrical field in a plasma is frozen on particles and remains on the ball lightning framework.For obtaining of observational parameters of ball lightning (Table III) is necessary that the initial electric field strength in an air is equal of order of $E_I \sim 10$ kV/cm.

Let us consider the mechanism of processes which create the high electric potential in an air and cause the above phenomena.Evidently they have the same nature as processes in a thunder-storm cloud ([24-28]) which lead to the origin of its electrical potential.After the formation of charged drops and pieces of an ice in the cloud in a result of different mechanismes takes place the charge separation because charged aerosol particles of size of 20-50 mcm are falling.In a result of this process the mean cloud of a thichness $L \approx 4$ km obtains the electric potential U = 30 - 50 MV, that corresponds to an average electric field strength approximately I0 kV/m (For comparing the average electric field strength near the Earth surface is equal I30 V/m ([29, 30]),the breakdown electric field strength for a dry air is equal 3 MV/m). It corresponds to an excess of the charge number density of one sign $U/(2\pi L^2) = 300\ e/cm^3$.Note that the typical number density of a plasma for a dry air is equal for these altitudes of order of $I0^3\ e/cm^3$.

It is essential for a creation of the cloud potential the presence of the large quantity of small particles with a radius a 60 nm.The attachment of air ions to these particles causes the decreasing of the atmosphere conductivity.For this reason an electrical current inside a cloud is small and does not discharge it.

Evidently the likewise mechanism determines the creation of an electrical potential of ball lightning.Let us considered it for two versions.In the first version at initial we have an atmospheric plasma with the charge number density more than $I0^9\ e/cm^3$.In this plasma takes place the charge separation as a result of a falling of large particles with sizes of order of I0 - I00 mcm.After this process small particles with sizes I - I0 nm are aggregated in a plasma and formate the charged framework of ball lightning.In the second version the charge separation in dust atmosphere proceeds by usual way as in a thunder-storm cloud,but the dimension of the dust region and the electrical potential are essential lesser than in usual cloud.After the cloud separation in a dust cloud near the interfaces arises a weak discharge in which takes place the formation of the ball lightning framework.

Let us make estimations for the first version.If a particle of a radius 20 mcm is introduced in a plasma of a dry atmosphere it obtains the charge approximately -I00 e due to difference of mobilities of positive and negative ions.If the number density of particles is large,its charge is lesser than the above one.A particle with a radius

20 mcm falls in an air with a velocity 5 cm/sec,i.e. the formation of an unipolar plasma durates of some seconds.Assume that the density of a particles material is equal 2 g/cm^3.Then the ratio ofthe charge of particles to their weight is equal 2 IO^{-IO} g/C.By usage of the average charge number density $IO^{-8.0 \pm 0.5}$ C/cm^3 (Table III) we obtain that the initial number density of a dust in an atmosphere must be equal $IO^{-I.I \pm 0.5}$ g/cm^3.This lowering estimation for the dust density is two order larger than the atmospheric density.Nonreality of such situation demands to refuse from the considered version.

The second version gives the following physical picture of given phenomenon.At the beginning we have a dust cloud with electrical potential of order of IOO kV.In a case of total analogy with the thunder-storm cloud this dust cloud has a size of order of IO m,really it can be essential lesser.In a gap between a cloud and a surface arises a weak gas discharge in which the ball lightning framework is formated due to presence in a gap of a corresponded material.Note that this picture is respected to the formation of ball lightning in an open air.Inside rooms can be existed other ways for the creation of the electrical potential of ball lightning.These ways are connected with interactions involving surface charges.

Thus electrical properties of ball lightning are determined by a character of its framework formation and the analysis of these properties gives the representation about mechanisms of the ball lightning formation.An electrical charge of ball lightning causes other phenomena.For example,the analysis shows ([8]) that the motion of charged endings of the ball lightning framework causes the radiation of ball lightning in the radiodiapason.It corresponds to the observational fact - an apperance of ball lightning causes strong distortions in radio receivers.

Understanding of electrical properties of ball lightning gives the possibilities for modelling of these properties.Interesting investigations of such type were fulfilled in Leningrad by V.Ja.Alexandrov and I.V.Podmoshenskij.The charged ball which was made from thin metallic wire and had a radius of some santimeters,was found in an air between two electrodes which were found one over the other with a distance between them I - 2 m.The ball wight was of order of the air one inside it,the electric field strength was of order of I kV/cm.Under an influence of this field a weak corona discharge arised near the ball wire.It provided the stability state of the ball in an air.This state could change under an influence of a displacing of surrounding objects,peoples etc. The main statement of authors of this models is thata weak corona discharge creats the stable state of the metallic ball in an external electric field.It contradicts to the above physical picture of electrical processes of ball lightning.At first,according to the analysis ([8]) the electric field of ball lightning can not cause a corona discharge near it.At second,electrical field strengths in this experiment are enough large and and correspond to a thunder-storm weather.In cases of observations of ball lightning inside rooms electrical fields must be essential smaller.But inspite of the above contradictions such method must take into account because it allows to check the validity of different conceptions on the base of simple models.

3.Radiative properties of ball lightning.

The luminosity of ball lightning is one of the essential properties of ball lightning Some radiative parameters of ball lightning are given in Table I.Further we shall analyze these data by kept of the general scheme.To obtain the representation about ball lightning as a radiative source,let us compare the radiative parameters of ball lightning with black body ones.Take a black body ball with the radius of middle ball lightning and determine its temperature,if this ball has the same radiative parameters as middle ball lightning.We obtain the temperature $T = I360 \quad 30$ K of the black body surface which creats the same light stream as middle ball lightning.Black body with the surface temperature T = I800 300 K has the same light efficiency as middle ball lightning. (Note that the light efficiencies of a match,of a candle,of an oil lamp are found in the range of the light efficiencies of ball lightning.) For black body, which has the light strem and the light efficiency of middle ball lightning;the ratio of the luminous surface square to the ball lightning surface square must be equal :

$$\xi = 10^{-1.8 \pm 1.2} \qquad (10)$$

i.e̊.ball lightning is transparency luminous body.

Thus on the base of this simple analysis one can obtain the following representation about ball lightning as a radiative source.At first,the temperature of radiating zones or radiating particles must be equal approximately 2000 K.This value is larger than give the above estimations,because energetical losses of ball lightning are determined by its gasdynamic and exceed the corresponded value of black body.At second, according to the above analysis the optical density of ball lightning as a radiative source is small.

Some representation about ball lightning as a radiative source gives the comparing of its radiative parameters with the flame ones.Assume that the flame occupies separate zones inside the considering volume.Believe that the flame radiation is created by small particles which size is smaller than the radiation wave-length and optical properties of these particles are similar to soot ones.At these conditions the optical depth of a system is small.Then the light stream is determined by a total mass of radiative particles in hot zones and does not depend on the volume of these zones.The total mass of a radiating substance,which creates the light stream of middle ball lightning (I 400 lum) is given in Table V at different flame temperatures.The necessary formulas were taken from book ([I5]).

The light efficiency η of Table V is the ratio of the light stream to the total radiation power.One can estimate more real meanings of this value by an usage of flame parameters.Approximately I0 % of the chemical energy of the candle flame,which temperature in a hot zone is I700 - I800 K, is spent to the radiation.Therefore the flame light efficiency is equal approximately 0.2 lum/Wt that coincides with the ball lightning light efficiency within accuracy limits.

Table V .

T,K	I600	I700	I800	I900	2000
m , 10^{-3}g *)	I.4	0.6I	0.27	0.I3	0.070
η , lum/Wt	0.85	I.5	2.5	3.9	5.8
τ , 10^{-3}sec **)	28	I2	5.4	2.6	I.4

*) The accuracy equals to $10^{\pm 0.2}$

**) The accuracy equals to $10^{\pm 0.5}$

Table V contents the life times τ of luminous particles in the flame,if its combustion essures the heat release power of middle ball lightning.Then we assume that as in the candle flame a fuel goes to hot zones in the state of small particles but not in the vapour state.These particles create the radiation and the heat release.Note for a comparison that the life time of fuel particles in the candle flame is of order of $3 \cdot 10^{-3}$ sec.According to this analysis the flame can model the luminosity of ball lightning.

For the further analysis of ball lightning as the radiative source let us use the variety of ball lightning colours (see Table I). It follows from this that chemical compositions of a glowing material can be different.Other conclusion is that observational colours,except white and red,can be created by the radiation of atoms and molecules of definite sorts.The observational diversity of colours can not be explained by a radiation of macroparticles or of a heated surface.

Assume that the ball lightning radiation is created by atoms or molecules in an air.Then it is essential that excited atoms and molecules are quenched efficiently in an air at the atmospheric pressure.In particular, Table VI contents values of the quenching rate constants for resonance excited states of alcaly metall atoms and parameters of these states. (The resonance excited states of atoms or molecules are the lowest electron excited states with a small radiative life time.) On the basis of this information one can determine the probability Z that the resonance excited atom will radiate,but will not be quencheq as a result of collisions with air molecules.This value is defined by formula :

$$Z = \left(\left\{ k(N_2) \left[N_2\right] + k(O_2) \left[O_2\right] \right\} \tau + I \right)^{-I} \qquad (II)$$

Here τ is the radiative lifetime of the considered excited state, $k(X_2)$ is the rate constant of quenching of this state as a result of a collision with the molecule X_2, $\left[X_2\right]$ is the number density of molecules X_2.

According to data of Table VI we have Z I. It means that excited atoms or molecules in an air are found in thermodynamical equilibrium with an atmospheric air. From this follows that the radiation of individual atoms or molecules,which are found in the hot space region,is determined by the temperature of this zone and does not depend on ways of the excited particles creating.This conclusion allows to choose the set of atoms and molecules which radiation can be responded for the ball lightning colour [8,9] Radiative transfers from resonance excited states of such atoms and molecules create photons in optical range of spectrum.

Table VI.

The parameters of resonance excited states of alcali metals atoms and the quenching rate constants for these states in an air at the atmospheric pressure and the temperature I800 K.

Atom, excited state	Excitation energy, eV	Radiative life time, 10^{-9} sec	Quenching rate constants, 10^{-10} cm^3/sec*) N_2	O_2	Probability of radiation, %
$Li(2^2P)$	I.85	27	5.6	--	I.I**)
$Na(3^2P)$	2.I0	I6	6.6	II	2.0
$K(4^2P)$	I.6I	27	5.0	I4	I.3
$Rb(5^2P_{\frac{1}{2}})$	I.56	28	5.0	I9	I.I
$Cs(6^2P_{\frac{1}{2}})$	I.39	3I	II	--	0.5**)

*) The average accuracy of these rate constants ([3I]) is about 20 %.

**) It is used that the ratio of quenching rate constants for molecules of oxygen and nitrogen is equal $3 \pm I$.

As radiative source ball lightning is similar to pyrotechnic illuminating compositions ([32-34]),which create the radiation of the definite colour.In both cases the radiation is a result of the chemical energy transformation and the luminosity of given colour is created by admixtures which are being in respectively small amount.Therefore pyrotechnic compositions can be used for modelling of ball lightning processes.Note that pyrotechnical compositions are more effective sources of the radiation comparing with ball lightning.Their light efficiency exceeds one order the ball lightning one.

Taking into account the above analogy let us are modelled chemical active substance of ball lightning by pyrotechnic compositions.By usage the specific energy content of pyrotechnic compositions 6 kJ/g [32] we obtain the density of an active substance in ball lightning $IO^{\pm 0.7}$ g/l. From this it follows,in particular,that combustion of an active substance can heat an air inside ball lightning up to temperatures of some tousends degrees.

Further we shall use the phenomenological model of ball lightning [8,9,I5,20] that follows from the many-sided analysis of this phenomenon.Ball lightning has a hard frameworkwhich is similar to an aerogel [20] and has the average weight of order of the air one.A small part of pores of this framework is occupied by an active substance.A heat release in a result of chemical reaction,involving an active substance, causes the heating of a surrounding space and creates a radiation.Our task is to analyse the character of this process.

Let us consider the simple model of propagation of a chemical reaction inside ball lightning after finishing this reaction for an individual element.The chemical reaction for one element of active substance leads to the heating of some space zone which is occupied by gaseous products of the reaction.If other element of the active substance is found in this heated zone, it reacts and therefore expands the heated volume.If there are new elements of the active substance in expanded heated volume, this process proceeds.Otherwise this process is stopped.

Thus model according to its nature is similar to the chain chemical reaction.For this reason further we shall use the mathematical apparat of the chain chemical reactions [35]. Introduce the probability ofthe chain continuation which is now the ratio of the volume of reaction products V_p after a reaction of considered element of an active substance to the part of the ball lightning volume V_o which respects to this element of an active substance.We have

$$V_0 = \frac{m_0}{m} V$$

where V is the total volume of ball lightning, m is the total mass of an active substance, m_0 is the mass of a considered elements.Further

$$V_p = \frac{m_0}{\rho_0} \cdot \frac{\rho_0}{\rho_a(T)} \cdot \frac{A_0}{A}$$

where ρ_0 is the initial density of an active substance, $\rho_a(T)$ is the air density at the final temperature, A_0 is the average atomic mass of air molecules, A is the average atomic mass of products of the reaction.The factor $\frac{\rho_0}{\rho_a} \cdot \frac{A_0}{A}$ takes into account the products expansion after the reaction.According to these formulas we have :

$$\alpha = \frac{V_p}{V_0} = \frac{m}{\rho_a(T) V} \cdot \frac{A_0}{A} \qquad (I2)$$

Using for an active substance of ball lightning the parameters of the pyrotechnic composition of yellow colour ([32]) (the specific energy content is 6 kJ/g, the density is 2 g/cm^3,the average atomic mass of products is equal 36)and assuming the final air temperature in the range I800 - 2500 K,we have on the base of formula (I2) :

$$\alpha = I.4\ I0^{\pm 0.4} \qquad (I3)$$

Note that within limits of this model one can pich out two regimes of this process. If $\alpha < I$ the prrocess ends after reactions of the limited numbers of individual elements which are of oreder of $(I - \alpha)^{-I}$. If $\alpha > I$ the process is self-accelerated and all elements take place in it.In the second regime the total active substance is used.But it is self-accelerated process whereas ball lightning is respectively quie- phenomenon.It demand the more careful analysis of the second regime.

Use the model of the chain chemical reactions ([35]) for the considered regime.Let us assume that on the first stage of the process products occupy the volume v_I , on the second stage - the volume v_2 etc. According to a definition of the parameter α we have $\alpha = v_2/v_I$.The process is finished on the n-stage,when the total volume of ball lightning is occupied,i.e. $V = \sum_{k=1}^{n} v_k$. Since $\alpha = v_k/v_{k-I}$ we have from this relation:

$$V = \frac{v_1 \alpha^n}{\alpha - 1} \qquad (I4)$$

It gives us the total number n of the stages :

$$n = \frac{\ln\left[\frac{(\alpha-1)V}{v_1}\right]}{\ln \alpha} \qquad (I5)$$

According to formulas (I4) and (I5) we have that if the value α - I is not small,the considered process has an explosive character and a typical luminosity time has the order of the time of combustion of an individual element.For this way the process develops,if an active substance is found inside ball lightning in the state of grains.To overcome the contradiction between observational data and the model consequence it is necessary to assume that an active substance is found inside the framework of ball lightning in the state of separated threads (but not grains). To justify such possibility we adduce the results of paper ([36]) ,in which was studied the structure of dyes (rodamin B and malachite green)absorbed by porous glass.It appears that the absorbed dye formates inside a porous glass a fractal structure with the fractal dimension D = I.74 $\pm$ 0.I2.With taken in consideration the threadlike structure of an active substance inside ball lightning the considering process includes itself alongside with a propagation of the wave of a chemical reaction along separated threads a transfer of this wave from one thread to others.

Let us make some estimations for a propagation of the wave of a chemical reaction and the luminosity along a separated thread of an active substance.Denote an initial radius of an active substance element through r_o.After the expansion products of a rection occupy a cylindrical volume of a radius R_o. We have :

$$\frac{R_o}{r_o} = \left[\frac{A}{A_o} \cdot \frac{\rho_o}{\rho_a(T)}\right]^{1/2} \qquad (16)$$

where A_0 is the average atomic mass of air molecules, A is the average atomic mass of reaction products, ρ_o is the initial density of an active substance, $\rho_a(T)$ is the air density at the final temperature. Using for an active substance parameters of the pyrotechnic composition of yellow colour ($A = 36$, $\rho_o = 2\ g/cm^3$) we obtain from formula (I6) :

$$\frac{R_o}{r_o} = 60\left(\frac{T}{1000}\right)^{\frac{1}{2}} \qquad (17)$$

where the temperature of reaction products is expressed in K.

The velocity of a wave propagation along the tread of an active substance is equal

$$v = v_0 \frac{R_o}{r_o}$$

where v_0 is the wave velocity in the massive pattern.For pyrotechnical compositions we have $v_0 = 0.1 - 1$ cm/sec. further we shall use these values.

For an estimation of an initial radius of an active substance element we use that th light efficiency of the luminosity of a hot gaseous volume depends on its radius.Lĕt us demand that the observational light efficiency of middle ball lightning coincides with a considered hot volume one.Make estimations for ball lightning of yellow colour which luminosity is created by the radiation of excited sodium atoms.We use a general scheme of the calculation which is given in papers [8,I5,20]. Assume that the sodium concentration in reaction products is the same as its middle concentration in the surface lauers of Earth ,i.e.is equal [37] 2.8 % .In these calculations the luminous volume is a cylinder of a radius R_0 and of a length $v\tau$, where τ is the time of a heat transfer Results of calculations are given in Table VII and are obtained for the light efficiency of the luminous volume 0.7 lum/Wt and for the wave velocity in the massive pattern I cm/sec.

Table VII.

T, 10^3 K	r_0, mcm	τ ,10^{-3}sec	R_0, mm	$\bar{n}$, 10^3	N_e, 10^{13}cm^{-3}
1.7	14	73	1.1	0.38	0.16
1.8	11	39	0.88	0.61	0.38
1.9	8.5	23	0.70	1.0	0.80
2.0	7.0	14	0.59	1.4	1.6
2.1	5.8	9.3	0.50	2.0	2.9
2.2	4.8	5.8	0.43	2.8	5.1
2.3	4.2	3.9	0.38	3.7	8.4
2.4	3.5	2.7	0.32	5.3	13
2.5	3.1	2.0	0.29	6.6	21
2.6	3.0	1.6	0.28	7.0	31
2.7	2.7	1.4	0.27	8.1	44

Let us give the phenomenological model of ball lightning on the base of these data with some numerical estimations.An active substance inside the framework of ball lightning is found in the state of separated threads or,more exactly,in the state of fractal clusters.The wave of a chemical reaction propagates along separated threads and is accompanied by the luminosity of hot zones.This wave can transfer to neighbouring threads Let us estimate the average number of threads n which participate simultaneously in the considered process.The power of the process on a separated thread is equal

$$p = \pi r_0^2 v Q \rho_0$$

where Q is the content of the specific chemical energy in an active substance, ρ_0 is the density of an active substance.Using the parameters of a pyrotechnic composition ($Q = 6$ kJ/g , $\rho_0 = 2\ g/cm^3$) we have for the average number of burning threads:

$$\bar{n} = P / p \qquad (I8)$$

where $P = 2$ kW is the power of middle ball lightning.Table VII contents values of $\bar{n}$ for middle ball lightning. These values are obtained on the base of formula (I8) and their accuracy is estimated by a factor $I0^{\pm I.2}$.

Note that in the considered case,when the wave propagates along separated threads, the process is not critical to the value α of formula (I2) and it is the same for $\alpha > I$ and $\alpha < I$. This process accelerates according to the law $\sim \exp(-t/\tau)$, where τ is the typical time of increasing of the theards numberwhich participate in the process.It depends on a length of a separated thread and on its inflammability. for middle ball lightning we have $\tau \sim I$ sec.

Thus on the basis of observational data and the above analysis we can give some description of ball lightning as a light source.Ball lightning is a light source of the middle brightness and has a small efficiency of a chemical energy transformation to a light energy.Note that the same order of value has the efficiency of matches and a candle.The ball lightning luminosity is created by small space regions which are heated up to temperatures close to 2000 K;This heating takes place due to chemical reaction of an active substance which is absorbed by the ball lightning framework and is found in the state of separated threads inside ball lightning.The waves of the chemical reaction propagate simultaneously along the large number of threads ($\sim I0^3$).It produces an im - pression of the volume luminosity.

Let us turn our attention to once more side of a considered system.In hot zones side by side with excitation and radiative processes take place ionization processes.Thus hot zones are a plasma and at the atmospheric pressure this plasma is found in the thermodynamical equilibrium.The degree of an ionization of such plasma is determined by low - ionization admixtures.Table VII contents numerical values of the electron number density in hot zones due to sodium admixture which content (2.7 %) is assumed to be the sodium content in the surface lauer of Earth ([37]). Apparently the number density of electrons in hot zones is respectively high.Therefore hot zones have a high conductivity that can influences on electrical properties of ball lightning and may be responded for electrical damages maked by ball lightning.As an ionization source ball lightning leaves behind it some regions with a high conductivity.During a thunder-storm weather it can cause the electrical breakdown.Besides ionization phenomena accelerate the discharge of ball lightning and decrease its lifetime.At the same time hot zones with a plasma are sources of electromagnetic radiation,in particular,radiowaves.

4.Conclusion.

The material of this paper reflects the contemporary state of the ball lightning studies.On the one hand we have a large information about ball lightning observations, we can give some statements of principle that are respected to its nature.On the other hand our representations about the nature of ball lightning are schematic,we have some contradictions at working out in detail of separated elements of this physical phenomenon and can not overcome their due to lack of a physical information.Thus we are found on the definite stage of ball lightning investigations and the true understanding of such state must allow to fulfil the scientific search for the decision of this problem by an optimum way.

Our expierency of the study of ball lightning indicates the complication of this many-sided phenomenon.Therefore this problem can not solve on the base of one successful idea.Unfortunately many scientists think of such way and they work in vain.For the success it is necessary to divide the problem on separated elements and to study each of these elements with taking into account the results of precusors.Just such investigations determine our contemporary representations respected to this problem.

The main aspects of the ball lightning problem are the following :

I.The structure of a substance in ball lightning.

2.The energy storage and the energy transformation.

3.Chemical processes.

4.Thermal processes.

5.Gasdynamical processes.

6.Electrical processes.

7.Radiative processes.

Just to last from the above processes is devoted this paper.Evidently the understanding of each aspect is found on the different stage.The most complicated aspects are reflected to points 3 and 4.

the studies methods must be provided the understanding of processes in ball lightning and give numerical values of its parameters.I believe that now the most productive among them is the modelling of separated processes.Examples of such types are the above modelling of ball lightning electrical processes on the basis of a ball of metallic filaments which is found in an external electric field,the modelling ball light ning by a metallic filament knot ,heated by a laser radiation,for the investigation the carrying power of ball lightning ([38]) on so on.

For modelling it is convenient to use patterns and objects which properties are similar to ball lightning ones.Such analog of ball lightning is aerogel which has the same structure as ball lightning.Analogues of ball lightning in accordance with the transformation of the chemical energy to the light energy are pyrotechnic illuminated compositions and chemical sources of the energy,as a candle,matches and so on.Side by side with the modelling the essential contribution in understanding of the nature of ball lightning can give the analysis of observational data,which must be fulfilled on the new level with the large statistics.

In conclusion we note that now we are found on the interesting stage of the ball lightning study whenthe foundation is laid for the next jump in the understanding of this problem.One can wait that in the near future we shall obtain on this way a lot of interesting.

References.

(I). J.R.McNally: Oak Ridge National Laboratory Report, ORNL - 3938,I966.

(2).W.D.Rayle:NASA Tech.Note,NASA-TN-D3I88,I966.

(3).W.D.Charman:Phys.Rep.,54,26I(I979).

(4).I.P.Stakhanov:About Physical Nature of Ball Lightning (Energoatomizdat,Moscow, I985; in Russian).

(5).S.Singer:The Nature of Ball Lightning (Plenum Press,New York,I97I).

(6).S.Singer:Lightning,edited by R.H.Golde (Acad.Press,London,I977),p.409.

(7).J.D.Barry:Ball Lightning and Bead Lightning (Plenum Press,New York,I980);

(8).B.M.Smirnov:Phys.Rep.,I52,I77(I987).

(9).B.M.Smirnov: Sov.Phys.Uspekhi,29,48I(I986).

(I0).Y.H.Ohtsuki,H.Ofuruton:Nuovo Cim.,C,I0,577(I987).

(II).B.M.Smirnov:Sov.Phys.Uspekhi,I8,636(I975).

(I2).V.Yä.Alexandrov,E;M;Golubjev,I;V;Podmoshenskij:Sov.Phys.Tech.Phys.,27,I22I(I982).

(I3).V.Ja.Alexandrov et.al.:Sov.phys.Tech.Phys.,27,527(I982).

(I4).S.R.Forrest,T.A.Witten:J.Phys.A,I2;LI09(I979).

(I5).B.M.Smirnov:Problem of Ball Lightning (Nauka,Moscow,I988;in Russian).

(I6).W.R.Lee:Lightning,edited by R.H.Golde (Acad.Press,London,I977),p.52I.

(I7). P.Meakin:Phys.Rev.Lett.,5I,III9(I983).

(I8).M.Kolb,R.Botet:Phys.Rev.Lett.,5I,II23(I983).

(I9).D.Weitz,M.Oliveria:Phys.Rev.Lett.,52,I433(I984).

(20).B.M.Smirnov:Sov.Phys.Uspekhi,2I,I33(I987).

(2I).N.A.Fuchs:The Mechanics of Aerosol (Macmillan,New York,I964).

(22).N.V.Krasnogorskaja:Electricity of Low Lauers of Atmosphere (Gidrometeoizdat, Leningrad,I972;in Russian).

(23).A.Hussin et al:J.Aerosol Sci.,I4,67I(I983).

(24).C.B.Moore,B.Vonnegut:Lightning,edited by R.H.Golde (Acad.Press,London,I977),p.5I.

(25).M;Brook,T.Ogawa:Lightning,edited by R.H.Golde (Acad.Press,London,I977),p.I9I.

(267.J.Latham:Quart.J.Roy.Meteorol.Soc.,I07,277(I98I).

(27).E.R.Jayaratne,C.P.R.Saunders,J.Hallett:Quart.J.Roy.Meteorol.Soc.,I09,609(I983).

(28).J.E.Dye et.al.:J.Geophys.Res.,9I,I23I(I986).

(29).B.J.Mason:The Physics of Clouds(Claredon Press,Oxford,I97I).

(30). H.Israel:Atmospheric Electricity (Keter Press Binding,Jerusalem,I973).

(3I).B.M.Smirnov:Excited Atoms(Energoatomizdat,Moscow,I982;in Russian).

(32).A.A.Shidlovskij:Principles of Pyrotechnics (Mashinostroenije,Moscow,I973;translation Publ.No. ADA 00I 859,I974).

(33).J;H;McLain : Pyrotechnics (Franklin Inst.Press.,Philadelphia,I980).

(34).T.Shimizu:Fireworks,the Art,Science and Technique (Maruzen Co.,Tokyo,I98I).

(35).V.N.Kondratjev,E.E.Nikitin:Kinetics and Mechanism of Gaseous Reaction (Nauka, Moscow,I974; in Russian).

(36).U.Even et.al.:Phys.Rev.Lett.,52,2I64(I984).

(37).I.Jackson:Landolt-Börnstein Data,Group V,Vol. 2a (Springer-Verlag,Berlin,I984), p.248.

(38).V.P.Krainov,G.P.Lebedev,A.O.Nazarjan,B.M.Smirnov:Sov.Phys.Tech.Phys.,3I,II20(I986)

THE CANDLE FLAME AS A MODEL OF BALL LIGHTNING

Smirnov B. M.
Institute of High Temperatures
Izhorskaja 13, Moscow 127412, USSR

Summary
There is an analysis of observational data of the ball lightning brightness. It is considered the model of ball lightning in which take place the same physical and chemical processes as in the candle flame. Parameters of the above model are represented. This model is compared with other ones.

There is a rich experience of ball lightning modelling in a laboratory which is reflected in books and reviews of Singer[1,2], Barry[3] and following books[4,5] on ball lightning. This experience showed that it is difficult to achieve a success by modelling of ball lightning as a whole. Ball lightning is a complicated and manysided phenomenon. Therefore for understanding of its nature it is necessary to model separated sides of this phenomenon and to study its analogues according to some properties.

I want to give some examples of such models and analogues with which I had dealings. It is modelling of ball lightning lift by a knot of a thin wire heated by a laser radiation.[6] Heat transfer processes from ball lightning to a surrounded air are modelled by an electric iron[7]. Conventional analogues of ball lightning are an aerogel and pyrotechnical compositions 8. An aerogel has the same structure as a ball lightning framework, pyrotechnical compositions are analogues of ball lightning according to the character of the chemical energy transformation to the luminous one. These examples are a part of ones which are respected to ball lightning and were extracted because they are close to the author. Each of these and other examples allow to answer the separated questions which are respected to the ball lightning nature. The sum of such models gives the possibility to describe the physical picture of the phenomenon considered and in the future on the base of such information one can study ball lightning as a whole in a laboratory conditions.

On this way it is essential the more profound study of different processes and phenomena in the plane of ball lightning. Glowing of a candle flame is one of such phenomenon which is close to ball lightning according to its nature. In both cases the chemical energy transforms to the luminous one and both phenomena are characterized by closed optical parameters. Therefore it is of interest the ball lightning model in which parameters of processes of the candle flame are used. This paper is devoted to an analysis of such models.

For a study of luminous properties of ball lightning the value of its brightness is a matter of principle and for this reason we shall consider this question in detail. The information about the distribution of ball lightning on the brightnesses according to Stakhanov[3)] and Grigorjev[9)] is given in Table 1. Let us treat this information by assuming that the distribution function of ball lightnings on the brightnesses is determined by the simple formula;

$$W(P) = \exp(-\frac{P}{P_0}) \tag{1}$$

where W(P) is the probability that the light stream of ball lightning is equal P, P_0 is the middle brightness of ball lightning.

According to formula (1) the probability $W(P_i, P_k)$ that the brightness of ball lightning is more than P_i and less than P_k is equal:

$$W(P_i, P_k) = \exp(-\frac{P_i}{P_0}) - \exp(-\frac{P_k}{P_0}) \tag{2}$$

Let us define the respectively error by the relation:

$$\varepsilon(P_0) = \sum_i \left[W(P_i, P_{i+1}) - W_{obs}(P_i, P_{i+1}) \right]^2 \tag{3}$$

where W is given by formula (2) and W_{obs} is taken from Table 1. Evidently the less is this value, the closer is the distribution (1) to the observational one. In Fig. 1 is represented the dependence $\varepsilon(P_0)$ which is constructed on the base of the Table 1 data. The optimal value P_0 which is expressed in units of equi-

Table 1
A Comparison of the Ball Lightning Brightness with the Electrical Lamp Brightness

The power of equivalent electrical lamp, Wt	The Stakhanov data [3]		The Grigorjev data [9]	
	The number of cases	The probability,%	The number of cases	The probability,%
< 10	55	9.2	89	6.7
10 - 20	83	13,9	103	7.8
20 - 50	109	18.3	209	15.8
50 - 100	140	25.5	314	23.8
100 - 200	150	25.1	376	28.5
200 - 500	39	6.5	230 (200 - 500 and > 500)	17.4
> 500	21	3.5		
Total	597	100	1321	100

valent electrical lamp power is equal 90 ± 20 Wt according to the Stakhanov data, 120 ± 18 Wt according to the Grigorjev data and 113 ± 16 Wt according to the summarized data. The above error ΔP is choosen of such way that $\mathcal{E}(P_0 \pm \Delta P) = 2\,\mathcal{E}(P_0)$.

Thus on the basis of summarized data of Stakhanov and Grigorjev we obtain for the middle light stream of ball lightning:

$$P_0 = 1600 \pm 200 \text{ lum.}$$

Remark that we taken into account only statistical error. The above value is found in an agreement with the precedings results[8,10] which were obtained on the basis of Stakhanov data.

For the analysis of the considered model we shall use the measurements of candle flame parameters which were fulfilled by

A. D. Khahaev and colaborators in Petrozavodsk State University. The results of these measurements consist of following. The radiation spectrum of a candle flame is continuous. It follows from this that the radiation is created by small particles in a flame-clusters and aerosols*). The temperature of these particles is equal 1750 ± 80K, the light stream is equal 10 lum (with accuracy 30 %), it is transformed 1.2mg stearin per secunda. Assuming the total combustion of the stearin we obtain from this for the light output of a candle flame 0.2 ± 0.1 lum/Wt. It is given in Fig. 2 the light output of chemical and other light sources. It showes that the light outputs of ball lightning and a candle are coincided in the limits of the accuracy. It defines our attention to a candle as a model of ball lightning.

The very essential fact is that the total evaporationg fuel in the cadle flame is found in the state of small particles. One can define the fuel mass in the flame by using the value of the luminous volume and by assuming the optimal ratio between an oxygen and a fuel in the flame. The light stream of a hot zone is proportional to the total mass of a hard particles in the volume of this zone. By using of soot optical parameters one can define the total mass of hard particles in the hot zone. These calculations give that the total mass of a fuel in the hot zone is the same as the total mass of luminous hard particles. Thus a fuel is found in the state of hard particles in the flame. The accuracy of this statement is defined by the accuracy of operations fulfilled and is estimated by 50 %.

Let us use parameters of the candle flame for modelling of energetical and optical properties of ball lightning. Assume that in ball lightning proceed the same processes as in the candle flame. Then ball lightning consists of a sum of luminous elements and each of these is characterized by candle flame parameters (T = 1750 ± 80K, the light output is 0.2 ± 0.1 lum/Wt). In this case the number of luminous elements in ball lightning is equal $10^{1.9 \pm 0.4}$.

*) This fact is known for a long time and was remarked in the Faraday lectures "The chemical history of a candle": "Any bright flame contains hard particles". New results are that one can accompany this statement by numerical data.

Assume that the chemical composition of a fuel influences on the flame temperature. Then the temperature of hot zones is a parameter of ball lightning. Parameters of ball lightning as a function of the hot zones temperature are represented in Table 2, where m is the total mass of a fuel which is found in hot zones in the state of hard particles. The value m is obtained from the condition that the total light stream from hot zones is equal to the light stream of middle ball lightning and for a material of luminous particles are used optical parameters of a soot. The value η' in Table 2 is the light output of hot zones, which is equal to the ratio of the light stream to the total power of a radiation. The time of a sojourn of fuel particles in hot zones τ is determined from the relation;

Table 2

T,K	1600	1700	1800	1900	2000
m , 10^{-3}g *)	1.4	0.61	0.27	0.13	0.07
η', lum/Wt	0.85	1.5	2.5	3.9	5.8
τ, 10^{-3}sec **)	28	12	5.4	2.6	1.4
r , mm **)	5.7	4.0	2.8	2.0	1.5

*) The accuracy $10^{\pm 0.2}$

**) The accuracy $10^{\pm 0.5}$

$$\mathcal{P} = \frac{m\ Q}{\tau} \tag{4}$$

where $\mathcal{P}$ is the power of middle ball lightning, Q is the heat capacity of a fuel.

Assuming that a hot zone is a cylinder of a radius r we have for the time t of a cylinder expansion due to an intrusion in it of a surrounded air:

$$t = \frac{r^2}{4\mathcal{D}} \tag{5}$$

where $\mathcal{D}$ is the diffusion coefficient of air molecules in an air. In Table 2 are represented the values of r, which are obtained from this formula with using $t = \tau$.

Thus on the basis of processes which take place in the candle flame we have the following model of ball lightning. At first a fuel is absorbed by a framework of ball lightning and is found in its pores. Evaporating by an action of the flame heat it is mixed with an air and is burning. The number of burning zones inside ball lightning is large and is determined by a size of a separated zone, which is of order of some mm, and by the oxygen concentration inside ball lightning. There are some connections between neighbouring zones. The development of one of this is accompanyed by decreasing of the oxygen concentration in a surrounded region and causes the quenching of neighbouring hot zones. Such connections determine the number of simultaneously luminous zones inside ball lightning.

Remark that this model differs from the model[5,8,11] in which chemical active substance is modelled by pyrotechnical compositions. Then the luminous process is represented as a sum of burning waves which spread along of individual threads of an active substance. In this case an active substance of ball lightning includes simultaneously a fuel and an oxidizer. The analog of such model is the combustion of pyrotechnical compositions and the analog of the above model is the flame which is formated during the combustion of organic substanses in an air. Each of these models deserve the attention and the following experimental investigations must give the selection between these models of ball lightning.

The above model provides mean parameters of ball lightning (the energy, the power, the light stream, the light output and so on). According to this model ball lightning has a hard framework which can be modelled by a rare aerogel. This framework absorbs a small amount of a fuel (for middle ball lightning less than 1g). The combustion of this fuel in an air oxygen creates some flame regions which provide the luminescence of ball lightning. In middle ball lightning is burnt of order of hundred candles. The above model is conventional for experimental modelling of processes in ball lightning. Its experimental study will allow to answer some questions about the nature of ball lightning.

References

1) S. Singer; The nature of ball lightning (Plenum Press, New York, 1971)
2) S. Singer; in: Lightning, ed. R. H. Golde (Acad. Press, London, 1977), p. 409
3) J. D. Barry; Ball lightning and bead lightning (Plenum Press, New York, 1980)
4) I. P. Stakhanov; About physical nature of ball lightning (Energoatomizdat, Moscow, 1985; in Russian)
5) B. M. Smirnov; The problem of ball lightning (Nauka, Moscow, 1988; in Russian)
6) V. P. Krainov, G. P. Lebedev, A. O. Nazarjan, B. M. Smirnov; Sov. Phys. Tech. Phys. 31 (1986) 1120
7) B. M. Smirnov; The enigma of ball lightning (Znanie, Moscow, 1987; in Russian)
8) B. M. Smirnov; Phys. Rep. 152 (1987) 177
9) A. I. Grigorjev; Chemistry of a plasma (Energoatmizdat, Moscow, 1989; in press)
10) B. M. Smirnov; Sov. Phys. Uspekhi 29 (1986) 481; 30 (1987) 420
11) B. M. Smirnov; Nuovo Cim. (1988), in press.

Figure Captions

Fig. 1 The dependence $\varepsilon(P_0)$. Solid line-Stakhanov data[3], dotted line-Grigorjev data[9], drawing line-the summarized data.

Fig. 2 The light output of light sources. Solid line-black body radiator, ■ -electrical lamp, × -pyrotechnical compositions, ○ -middle ball lightning, □ -candle flame.

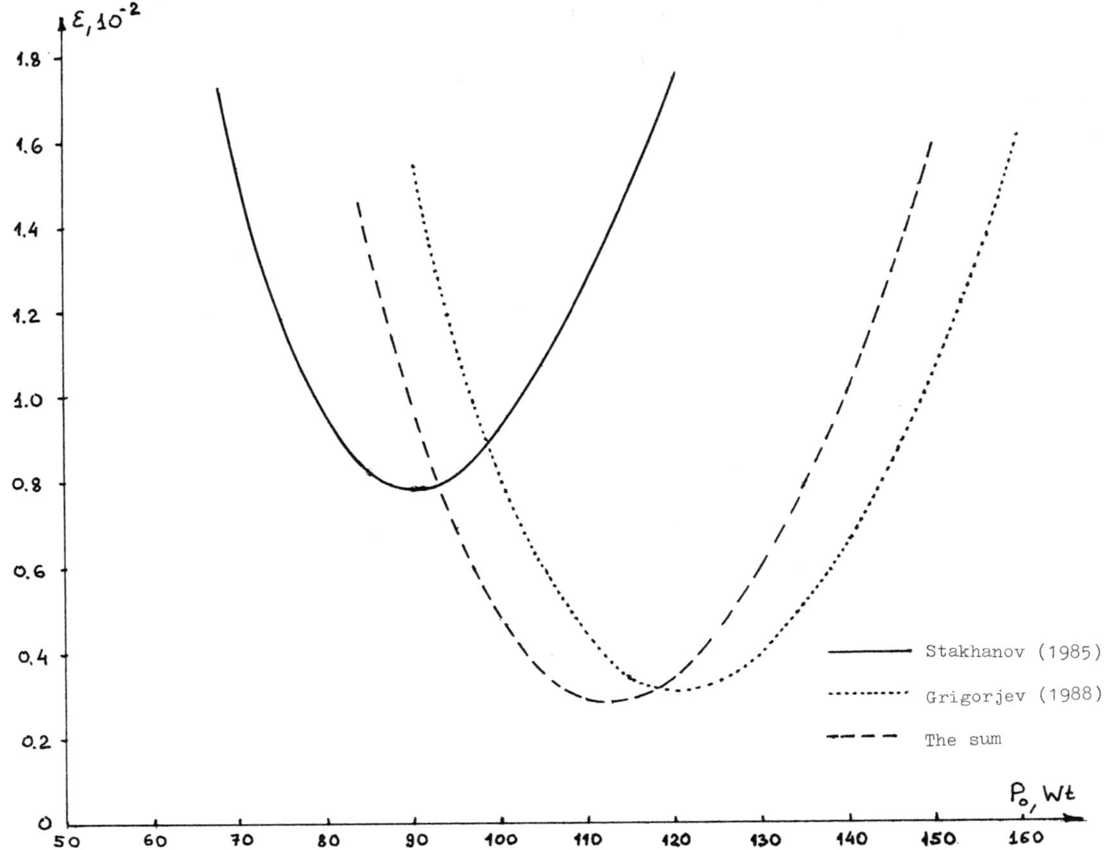
ε, 10⁻²
1.8
1.6
1.4
1.2
1.0
0.8
0.6
0.4
0.2
0
50
60
70
80
90
100
110
120
130
140
150
160
P₀, Wt
Stakhanov (1985)
Grigorjev (1988)
The sum

Fig.1

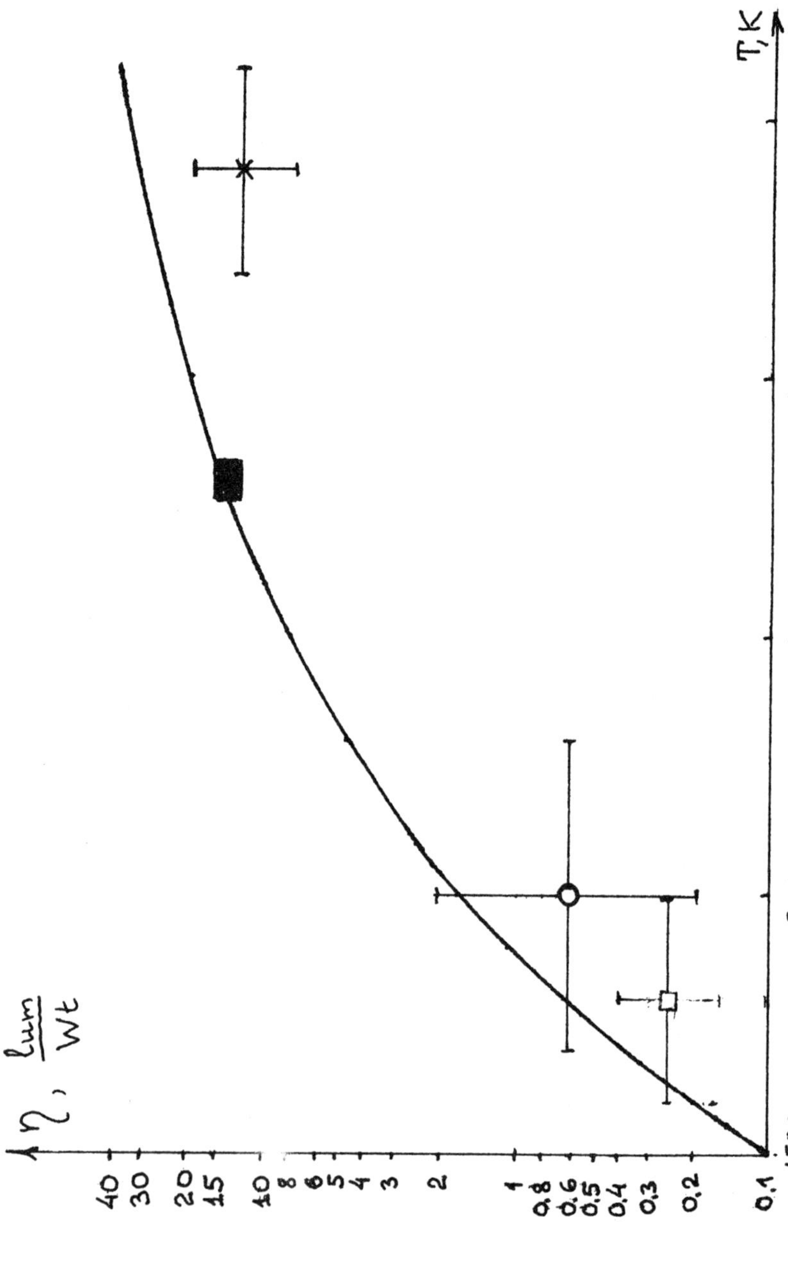
η, lum/Wt
40
30
20
15
10
8
6
5
4
3
2
1
0,8
0,6
0,5
0,4
0,3
0,2
0,1
1500
2000
2500
3000
3500
T,K

Fig. 2

A Model for Ball Lightning

Hiroshi Yamamoto
Department of Applied Physics
Fukui University, 910 Fukui, Japan

Abstract

Properties of a kind of ball lightning are summarized briefly. A model for this ball lightning is proposed. It is asserted that the ball lightning is a self-confined plasma. No external field or no aerosol to confine it is assumed. The above properties are well explained by this model. An experimental method is proposed to reproduce the ball lightning in a laboratory.

1. Introduction

Since the nineteenth century the observations of ball lightning were reported by the eyewitnesses fairly often, but they did not attract much attention of physicists[1]. The reason is probably that the phenomenon of ball lightning occurred very seldom and was observed by very few people who were not scientists. Furthermore the phenomenon is very curious and difficult to explain physically or to reproduce by experiment. They were sometimes told in connection with a man's soul in Japan.

As is also seen in this proceedings, there seems to exist many different kinds of ball lightning. One is, for example, very destructive and another is very gentle or quiet. The colors are also very rich (red, orange, green, blue, mixed and so on). Some ball lightning has completely different structure from the other. It seems impossible to explain all the properties of ball lightning by a single model. Therefore we must make clear what kind of ball lightning we are considering. In the following section the properties of the ball lightning which is dealt with in this paper are summarized. In section 3 a model for this ball lightning is proposed and the properties are explained by this model. In section 4 we suppose in what manner the ball lightning might be produced in nature. Lastly we propose an experiment to reproduce the ball lightning in a laboratory. In the final section I would like to mention the motivation I began to consider this problem and what I felt in this symposium.

2. Properties of a kind of ball lightning

According to the preprint of G. Egely[2] the eyewitnesses describe their experience very vividly. So we can figure the ball lightning to ourselves easily. We notice that there are many kinds of ball lightning in the reports. We summarize here the properties of a kind of ball lightning which we are intending to analyze. The most interesting and striking feature of the ball lightning to us is that it comes into a house and goes out through an open window or door. It appears mostly with thunders. Hence it must be an electromagnetic phenomenon. Though there seems to be no external electric or magnetic field to confine it, it is very stable, because the lifetime is more than a few seconds. It moves very freely with velocity about a few meters per second. It does scarcely collide with anything, for example, no collision with walls, floors, ceilings, furnitures or humans. Some ball lightning passed through near persons. At that time they did not feel neither electricity nor heat at all but a very gentle breeze. It seems to be a mass of gas and moves through the space with air or pushing it away. The size of ball lightning of this kind is not very large but about 10-50 cm in diameter. The light from it is not dazzling, but one can gaze at it. The color is usually red. The persons did not hear any noise when the ball lightning passed near by.

3. Model

From these properties we see at once that the ball lightning is neither burning material, nor chemical reaction, nor charged object at all. The ball lightning of this kind is nothing but a plasma in the air. The plasma consists of electrons and ionized molecules of the air. Therefore it is neutral as a whole. The usual plasma in the laboratories radiates no heat or makes no noise at all. It is not dazzling usually. The important point is to explain what mechanism makes it stable and movable in free space. The free space here means that there exists no external electric or magnetic field to confine it or to make it stable. As such a mechanism I propose a model of plasma in which electrons move like in Fig.1. The plasma should confine it by itself and make a closed system. As is well known, electrons are trapped by magnetic fields and move usually along the magnetic fields. Since there is no external magnetic field, the electrons should produce the confining magnetic field by themselves. The electric current inside of the plasma A confines the electrons outsides B and the electrons outsides B confine the electrons insides A. This makes the plasma stable.

To see how stable it is, we must analyze it mathematically, or probably by computer simulation. Because I did not do it yet, I cannot say much about it, but it depends of course on the size of plasma, the density and velocity of electrons, the pressure, temperature and components of the air and so on. As the reason of radiation two kinds of mechanism can be considered. One is by the acceleration of electrons. The electrons make circular motion and thus emit the synchrotron radiation. The wave length of this light is given by

$$\frac{h}{p} = 2.35*R/E^3$$

where $\frac{h}{p}$ is in Angstrom[Å], R is the radius of circular motion in meter[m] and E is the energy of electron in giga-electron volt[GeV]. This length is too short and the corresponding light is invisible. The other mechanism is by ionization of molecules through the collisions of electrons. A free electron falls again into a vacant energy level of an ionized molecule and at the same time it radiates a photon with corresponding energy. This is the mechanism of radiation of usual plasma. The color of usual plasma depends mainly upon the sort of gas of atmosphere in which the plasma occurs. The colors of plasma in different kinds of gas are shown in Table 1[3]. The plasma in this case is in very low density atmosphere. However, if the mechanism of luminescence is same as in the case of high density, the color in the atmospheric pressure must not be very different from the above. P. M. Koloc said in the symposium that the spectrum of plasma in higher density atmosphere becomes broader. According to the preprint by G. Egely[2] the distribution of colors of ball lightning reported by the eyewitnesses is as in Fig.2. It means, our description agrees very much with this distribution. The atmosphere of the ball lightning is mainly the air. Especially the ball lightning of pink color means that the atmosphere is of high humidity.

According to our model of ball lightning it occurs in a certain domain of the air and the air inside of it is in a special plasma state. It forms a closed stable state and is independent of the outside. Therefore the ball lightning moves with air. The motive force is probably a flow of the air but strong wind may break the stable state. Hence the ball lightning of this kind moves very slowly and cannot penetrate any dense material, for example, a closed window or door, since the air cannot go through it. Because the ball lightning is neutral, it is neither attractive nor repulsive to charged matter or conductive metals. These properties agree with the observations by the persons, whose neighborhood the ball lightning passed by. They felt only soft movement of the air. In this way many properties of ball lightning are explained by this model.

4. Production of ball lightning

Next problem is to explain by what mechanism the ball lightning is produced in nature. It is a very difficult problem, but I tentatively consider as follows (Fig.3). At a thunderbolt a big current flows between the sky and the earth. The electric current induces a magnetic field around it and electrons are trapped by this field. The trapped electrons induce also magnetic fields by which other electrons are trapped. These electrons induce again magnetic fields by which the original electrons are trapped again and thus the system is closed to finish a ball lightning. Of course no one knows how to reproduce it in a laboratory by experiment[4]. However, if one sets a ring circuit with a direct current into usual plasma, one will see that a kind of dense plasma arises around the circuit (Fig.4). It is of course not exactly a ball lightning, but the form is akin to it. We might be able to investigate it and predict something by this plasma.

At the end of this paper I would like to propose a possible experiment to produce a ball lightning in a laboratory. This is of course just a conjecture. First prepare an atmosphere filled with ionized molecules of air. This is probably possible by projecting electron shower to the air or by radiating a light beam of a suitable frequency to it, which is resonantly absorbed by the air. Then make a discharge between two electric poles that are placed in the atmosphere. If the situation is suitable enough, we might expect ball lightning to occur.

Acknowledgment and some comments

First of all I should acknowledge Prof. Y.H.Ohtsuki for giving me an occasion to have a talk in this symposium. As I mentioned there, I am not an expert of the ball lightning. I am just a theoretical particle physicist. Recently I was happy to have the motivation to study the ball lightning. I receive preprints periodically from Hungarian Academy of Science, Central Research Institute for Physics, Budapest. Last year a preprint by Prof. G. Egely[2] was sent to me. I was very much interested in it and read it when I was tired from doing my daily occupation. While I was reading it, an idea came across my mind. I wrote a short note and sent it to some people who seemed to me to be interested in it. After that I found many papers on ball lightning published recently in the computer database and asked the authors of the papers to send me the reprints. I received some of them with the kind compliments of the authors[5]. I was very pleased at them and am very grateful to the authors. From these papers I have learned a lot, for example, the history of observation and research of ball lightning is very old for about 150 years. Of course it should have been from the ancient times. Nevertheless the number of scientists who are interested in it is not very large still at the present day. The usual scientists do not know much about ball lightning, because they are absorbed in the fashionable topics. Some japanese scientists consider that the fire ball (=ball lightning) would come from a grave and be seen in a cemetery. Hence they regard it as a problem of psychics and no problem of natural science. At most they consider that it is burning phosphorus and has nothing to do with lightning. However I noticed that when they were explained the actual properties of ball lightning, they were very much interested in it, discussed about it and proposed their own models. Hence I think, if the real feature of ball lightning is known universally in future, many scientists will begin to study it and gather to the symposium of ball lightning.

In the present symposium many observations of the eyewitnesses were reported. We recognized that there are many categories of ball lightning. We must not discuss them on equal terms. Every one of the participants has his own image of ball lightning which is sometimes completely different from others. The image comes from what property of ball lightning he is mostly interested in. Therefore our discussions were occasionally on different wavelengths. The reports of eyewitnesses were often dealt with statistically, but things of different categories were mixed up in a category. Consequently many important informations are buried under the statistics. In this sense the direct reports of eyewitnesses should be taken good care of. We must collect them and publish them in their original live forms. What one learns from the collection or which statistics one extracts from it should be left to the readers. Then we may expect that the observations of ball lightning will be classified reasonably and each kind of ball lightning will have consistent properties. In conclusion I would like to express my sincere thanks to all the participants for their kindness extended to me.

References

1. S. Singer, "The Nature of Ball Lightning", New York, Plenum Press, 1971.

   J. D. Barry, "Ball Lightning and Bead Lightning", New York, Plenum Press, 1980.

2. G. Egely, "Hungarian Ball Lightning Observations", Preprint of Central Research Institute for Physics, Budapest, KFKI-1987-10/D.

3. T. Idehara, Private communication. He is one of my colleagues at Fukui University and an experimental plasma physicist. I am grateful to him for his interest shown in this problem.

4. I was very much interested in a photograph taken by R. K. Golka in a Locomotive experiment, which was shown in the symposium. It seemed to me very resemble to the ball lightning that we are considering here.

5. J. D. Barry, Frequency of ball lightning reports, J. Geophys. Res. 84, 308-310 (1979); On the energy density and forms of ball lightning, J. Geophys. Res. 85, 4111-4114 (1980).

   P. Brovetto, V. Maxia and G. Bussetti, On the nature of ball lightning, J. atmos. terr. Phys. 38, 921-934 (1976); Some remarks about a ball lightning model, J. atmos. terr. Phys. 44, 925-930 (1982).

   W. N. Charman, Ball lightning, Phys. Report 54, 261-306

(1978); Ball lightning: The unsolved problem, Weather 37,66-75 (1982).

G. C. Dijkhuis, Thermonuclear energy from ball lightning, Proc. 14th Intersociety Energy Conversion Engineering Conf., Boston (1979); A model for ball lightning, Nature 284,150-151 (1980); Threshold current for fire ball generation, J. Appl. Phys. 53, 3516-3519 (1982).

S. Singer, Ball lightning, The persistent enigma, Naturwissenschaften 67, 332-337 (1980).

D. R. Tompkins, Jr., Would shower cores or relativistic monopoles produce straight lightning ?, Phys. Rev. D4, 1268-1274 (1971); D. R. Tompkins Jr. and P. F. Rodney, Possible photographic observations of ball lightning, Nuovo Cimento 3c, 200-206 (1980).

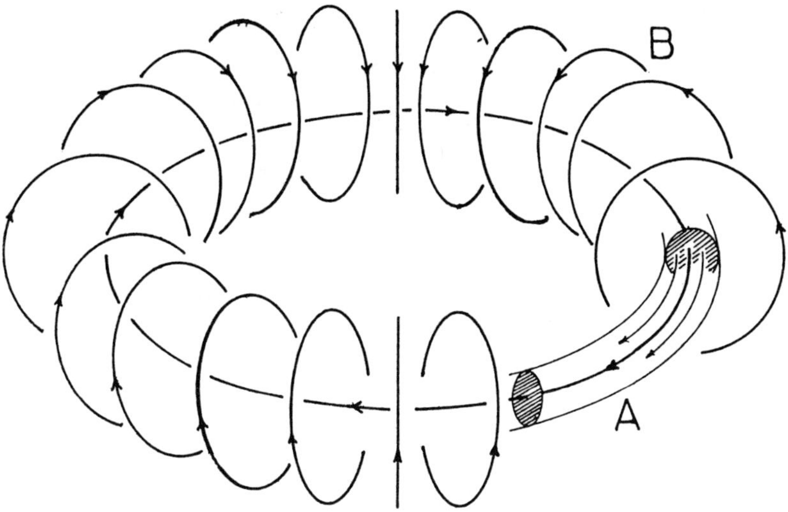

Fig. 1. Electron current in the plasma of ball lightning.
A: inside current,
B: outside current.

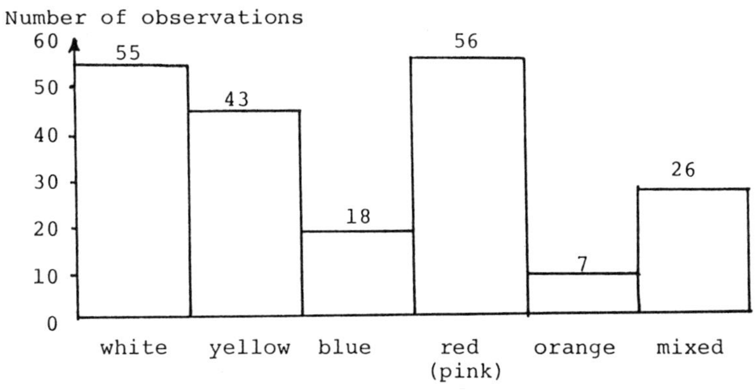

Fig. 2. Distribution of the reports as a function of the colour of the ball lightning. (taken from the preprint of G. Egely)

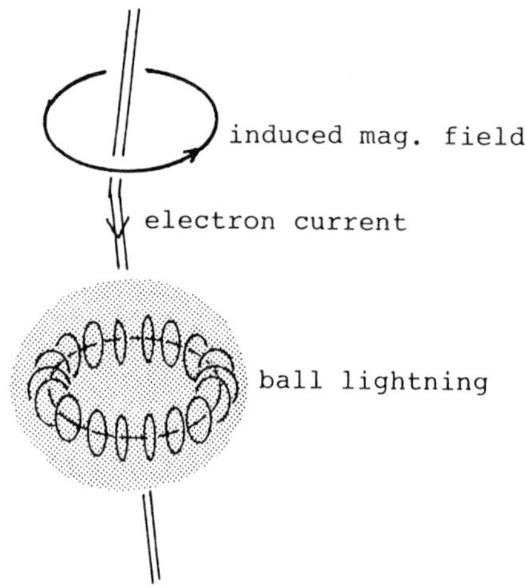

Fig. 3. Formation of ball lightning at a thunderbolt.

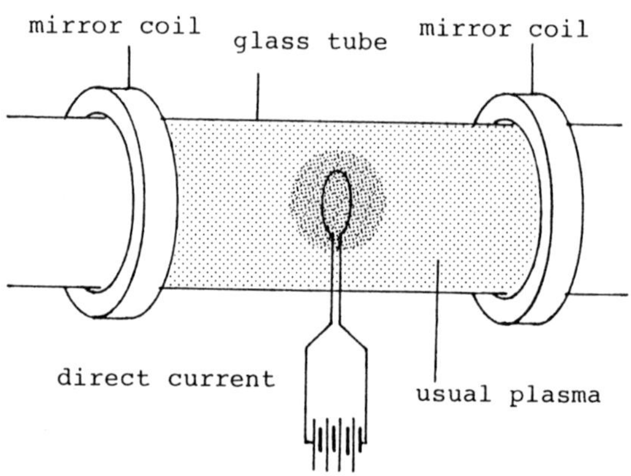

Fig. 4. Quasi ball lightning in usual plasma.

Table 1. Colors of plasma in different kinds of gas.

gas	He	N_2	air	H_2O	CO_2
color	orange	lurid red	red	pink	white

THE HIGH-TEMPERATURE PHYSICO-CHEMICAL PROCESSES IN THE LIGHTNING STORM ATMOSPHERE
(A Physico-Chemical Model of Ball Lightning)

Georgy P. Gladyshev
Institute of Chemical Physics
USSR Acad. Sci., Moscow

Numerous research related to physico-chemical transformations in atmosphere discuss specificity of the processes going under effect of the solar radiation, electric discharges and other factors.[1-12] However, a range of phenomena have not found a theoretical explanation based on exact calculations or experimental data yet.

This work is focused on studies of the phenomena related to the high-temperature chemical processes happening in the intensive atmospheric fields, i. e., basically, to the ball lightning phenomenon*. The key objective of this study is to demonstrate that the ball lightning can be a diffusional flame maintained by the atmospheric d. c. currents.

The thermodynamic and kinetic analysis results[13-14] make it possible to single out the basic chemical reactions going on in the lightning storm atmosphere as well as the high-temperature ionization processes shaping the ball lightning phenomenon.

Firstly, let's discuss the specific features of the physico-chemical transformations of the lightning storm air in a wide temperature range. In the lightning storm atmosphere nitrogen oxides, ozone and other "combustible" components start to accumulate reaching concentrations in excess of the normal levels (sometimes by several tenfolds). Thus, nitrogen oxide is syn-

---

* An indentified flying object observed by the author, in March, 1980, 4p. m., at a site near Moscow in low continuous cloudiness (temperature about 0°C). The object was observed over the forest, at distance some 70-100m away, at altitude of 50-80m. The object remained motionless and was observed for over 15 minutes. It hardly looks like a typical ball lightning. However, the object image resembles quite much the ball lightning photograph taken by Charman W. N. (New Scientist, 56, 1972, p. 632).

thesized through the following process

$$N_2 + O_2 \rightleftharpoons 2NO - q\ (43.5\ Kcal) \quad (1)$$

where: the reaction heat effect ($-q \equiv \Delta H^o$) relates to standard conditions. Having entered the low-temperature zone, NO is "tempered"[15,16] and can react with oxygen and ozon as in the below equations

$$2\ NO + O_2 \rightleftharpoons 2\ NO_2 + q\ (27\ Kcal) \quad (2)$$

$$NO + O_3 \rightleftharpoons NO_2 + O_2 + q\ (47.5\ Kcal) \quad (3)$$

Under the lightning storm conditions the processes described by the cumulative equation of the below type become significant too

$$2\ NO_2 + O_3 \rightleftharpoons N_2O_5 + O_2 + q\ (64\ Kcal) \quad (4)$$

The created N_2O_5 dissociates in rain drops and fog particles in the following way

$$N_2O_5 \rightleftharpoons NO_2^+ + NO_3^- - q \quad (5)$$

It should be noted that N_2O_3, also a product of the cumulative process, is similarly unstable even at low temperatures and in the liquid phase decomposes into NO^+ and NO_2^-. In addition, there are other known processes of the type

$$2\ NO_2 \rightleftharpoons NO^+ + NO_3^- - q \quad (6)$$

$$H_2O + HNO_3 \rightleftharpoons H_3O^+ + NO_3^- - q \quad (7)$$

increasing the liquid phase conductivity by several orders. The reactions of N_2O_5 and N_2O_3 decomposition in the gaseous phase under the appropriate conditions can generate some amounts of ions as well.

Considering the ionization potentials and the values of affinity to electron and proton of the particles identifiable in the atmosphere, one can state that at temperatures of 300-500K the atmospheric air contains NO^+, NO_2^+, O_2^+, H_3O^+, NO_3^-, NO_2^-, O_3^-, CO_3^- and some other ions in relatively increased concentrations. However, it should be noted, that due to the kinetic and thermodynamic limitations as well as to the specificity of the nitrogen based oxygen compositions chemistry, the ion composition in the atmospheric air under effect of temperature changes can vary in the real environment in a complex manner (e. g. reaction 1 in the above range has some kinetic limitations; NO_2 can withstand temperatures practically up to 900K, whereas thermal decomposition of N_2O is noticeable at 800K and above, etc,).

The most significant role in the charged particles generation in the atmosphere is probably played by the relatively advantageous, in terms of thermodynamics, ionization processes involving electron transfer. Thus, e. g. interactions

$$NO + NO_3 \rightleftharpoons NO^+ + NO_3^- - 7.5 \quad 5.4 \quad eV \tag{8}$$

$$NO_2 + NO \rightleftharpoons NO^+ + NO_2^- - 6.2 \quad 5.3 \quad eV \tag{9}$$

$$NO_2 + NO_3 \rightleftharpoons NO_2^+ + NO_3^- - 8.0 \quad 5.9 \quad eV \tag{10}$$

$$NO + O_3 \rightleftharpoons NO^+ + O_3^- - 7.4 \ eV \tag{11}$$

can intensively proceed in the atmosphere including the lightning discharge area, and roles of different reactions can vary depending on the temperatures (at the right hand side of equations 8-10 the upper possible values of the processes' heat are indicated, whereas the lower limit values are given in brackets).

The mentioned above ions are easily registered in the atmosphere and ionospheres of planets[9-17]. However, it has been often mentioned that the ion concentrations are significantly higher than those calculated by the Saha-Boltzman equation[18-20]

$$\frac{n_i n_e}{n_\mu} = \frac{g_i g_e}{g_\mu} \cdot \left(\frac{m \cdot M_i}{M_\mu}\right)^{3/2} \cdot \frac{(2\pi kT)^{3/2}}{h^3} \exp\left(-\frac{I}{kT}\right) \tag{I}$$

because the equation was obtained assuming that the thermal ionization of atoms and molecules (A) in the system goes in the spontaneous way only

$$A \rightleftharpoons A^+ + e^- \tag{12}$$

and the electric ionization and other ionization processes on surfaces of solid particles always present in the atmosphere are not accounted for. In relation (I) the following symbols are used: n_i, n_e, n_μ - concentrations of ions, electrons and neutral atoms or molecules; g - corresponding statistic weights; m, M_i, M_μ - masses of the particles; I - ionization potential.

For example, comparison of the experimental values of ion

concentrations in the lightning storm atmosphere (containing fivefold excess of N_2O_5, NO, H_2, NH_3, CH_4) at 1000K as well as those in the corresponding flame zones, against the values calculated using the Saha equation assuming even the abnormally low mean value of I = 6 eV, shows that the experimental data exceed the calculated values by several orders.

On the other hand, the experimentally defined values of n_i in the lightning storm atmosphere correlate well with the estimates obtained by extrapolation of the experimental data on transformation of the oxygen compositions of nitrogen in liquid media with a low dielectric permittivity where effective dissociation of the above compositions is observed with generation of NO^+, NO_2^+, H_3O^+, NO_2^- ions.[21-23] This is yet another evidence of the solvatational and other effects role in dissociation in the real atmospheric environment.

Thus, we come to the conclusion that calculations by Saha assuming only process (12) going on at temperatures about 1000K for the real atmosphere give significantly underestimated values of n_i.

However, the theoretical values of n_i determined using formula (I) for the air at temperatures 2000-2500K should have a better correlation with the experimental data. The fact is that under such temperature the type (12) processes become dominating, because under the equilibrium conditions the air is practically a mixture of several gases: O_2, N_2, NO, Ar, H_2O with higher concentrations as compared to other components (e. g. CO_2, H_2), and the thermal ionization proceeds using only one component – NO (with a relatively small ionization potential of 9.267 eV). In addition, under such conditions the ionization involving solid particles in the real atmosphere should be small.

Let's now discuss the ball lightning model in general. An electric lightning discharge or a volumetric charge discharge in the atmosphere creates in some point in space an elevated ion concentration and a high temperature, i. e. the conditions required for the ball lightning flame initiation. The flame is of ball (hollow sphere) shape and is maintained by dc atmospheric currents.

Assume that in the relatively cold zones of the sphere the type (2-3) processes are maintained by diffusion of the "fuel" and oxidants - NO, H_2, (NO_2), O_2, (O_3) and others.

It can be easily shown that even at the abnormally high concentrations of the combustible admixtures (about 1 %) and the process mean heat of 35-40Kcal/mole the heat up can not raise temperature even to 600-700K. The real content of the combustible products (including CH_4) in the lightning storm atmoshere in no case can exceed 0.0003 % (in volume). Thus, the energy of the discussed chemical processes is by several levels below the level required for the ball lightning heat up.

Moreover, consideration of a stationary diffusional burning of the fire ball in a stable medium, which in a simplified model can be described by a simple heat transfer and diffusion equations

$$\lambda\left(\frac{d^2T}{dr^2} + \frac{2}{r} \cdot \frac{dt}{dr}\right) + q\phi(T, 0) = 0 \qquad \text{(II)}$$

$$\mathcal{D}\left(\frac{d^2C}{dr^2} + \frac{2}{r} \cdot \frac{dC}{dr}\right) - \phi(T, C) = 0 \qquad \text{(III)}$$

with reasonable assumptions (on small areas of the on-surface transformation zone, etc.) it easy to establish that this model is unstable against radial excitations.[24] In equations (II, III): T - temperature; C - concentration of combustible matter; ϕ(T, C) - chemical reaction rate; r - radial coordinate.

Thus, we understand that is hardly possible to create a purely "chemical" model of the phenomenon without an external energy source.

If some burning gaseous sphere (a ball) containing ions of different polarity is in the atmospheric electric field, there should be separation of the on-surface (and volumetric) charges effected in the sphere by the field (as well as by the thermo-diffusional and other effects though to an insignificant degree). The charge of such a burning sphere (a ball) and the field intensity on its surface can be estimated using the known model of field and model of charge distribution on a conductive sphere in an external uniform electric field.

Assume, that a sphere of radius r_a is positioned in a uniform external electric field of intensity $\vec{E}_0$. It is easy to show that the charges on the sphere's surface are distributed by the law

$$\sigma = \varepsilon_0 E = 3\varepsilon_0 E \cos\theta \tag{IV}$$

where: σ - on-surface charge density; θ - polar angle; ε_0 - dielectric constant. Such a distribution of charges is characterized by the dipole moment

$$\vec{P} = \int \vec{r}\sigma \, dS = 4\pi \varepsilon_0 r_a^3 E_0 \vec{e}_z \tag{V}$$

where: dS - element of area; r - radial coordinate; $\vec{e}_z$ - unit vector along z axis in the E_0 field direction. The electric field beyond the sphere is defined as

$$\vec{E} = \vec{E}_0 + \frac{3\vec{r}(\vec{P}\vec{r}) - \vec{P}r^2}{4\pi \varepsilon_0 r^5} = \vec{E}_0 - \frac{\vec{P}}{4\pi \varepsilon_0 r^3} + \frac{3\vec{r}zP}{4\pi \varepsilon_0 r^5} \tag{VI}$$

It should be noted that the ball's own electrostatic field intensity, the charge of which is uniformly distributed over the ball surface, coincides beyond the ball limits with the point charge intensity, q, positioned in the ball center. It is hard to say to what degree of approximation the ball lightning can be treated as a burning sphere. It may well be closer to aburning ball charged uniformly over all volume. However, at the phenomenon principle model creation it does not matter which of the cases is realized in nature (it is not excluded that the both ones). So, let's estimate density of the conductivity current, $\vec{i}$, for the "droplet" case

$$\vec{i} = \vec{i}_+ + \vec{i}_- = (\lambda_+ + \lambda_-)\vec{E} = -\sum_i \lambda_i \nabla \varphi \tag{VII}$$

where: λ_+, λ_- - conductivities of the positive and negative ions; φ - potential ($\vec{E} = -\nabla\varphi$).

Using relation (VII) let's estimate density of the conductivity current for the realistically possible values of the ion concentrations, n_i, potential gradient $d\varphi/dr(\nabla\varphi)$ in the lightning storm atmosphere at temperature of 2500K. Selection of this temperature value is fairly well substantiated because the ball lightning temperature varies in a wide range and as a rule should reach this selected value.

Assuming $n_i = 10^9$ ion/cm^3, ion mobility $u_i = 50$cm^2/s•V, We obtain $\lambda = \sum_i n_i u_i e = 7 \cdot 10^3\ s^{-1}$, where: e - charge of electron ($4.8 \cdot 10^{-10}$ CGSEq). Further, taking into consideration that in the pre-discharge instance $|d\varphi/dr|$ of the linear lightning can be equal to 10^4 V/cm^2, the value of $\vec{i}$ in the "droplet" is 10^{-4} A/cm^2. Note, that the same density of current is obtained at $n_i = 10^{10}$ ion/cm^3 and $|d\varphi/dr| = 10^3$ V/cm, etc.

The obtained value of current is quite adequate to observe the diffusion-limited endothermic synthesis (1) and to maintain the gaseous ball (sphere) at a temperature of many hundred degrees. The obtained estimates correlate well with the theoretical analysis and other calculation data. It can be shown that the model of the diffusionally controlled reaction in the sphere (burning, NO synthesis, etc.) maintained by dc current of power below the break down threshold is similar to "burning" of the spherically symmetrical optical charge[14,20] and is stable against radial excitations.

Inflow of current heats up the sphere providing for NO synthesis as it has been mentioned above. Additional quantities of nitrogen oxide can be generated by the electric synthesis associated with the reactions of the below type

$$O_2 + e \rightarrow O + O^* + e \tag{13}$$

$$N_2 + e \rightarrow N_2^* + e, \text{ and other} \tag{14}$$

Synthesis of O_3, NO_2 and other products[25] can go in the relatively cold areas of the sphere and its adjacent environment.

Reaction (1) has been studied quite well; ΔH^o and ΔS^o of the process are known. Hence, from equation

$$\frac{\ln K_p}{(1/T)} = \frac{-\Delta H^o}{R}$$

it is easy to calculate the reaction coefficient K_p and, correspondingly, NO concentration in the equilibrium mixture at the given temperature. Further, by putting the n_{NO} concentration value into the Saha equation (I) it is possible to estimate ionization level using relation[20]

$$\lg \frac{(n_{NO^+})^2}{n_{NO}} = \lg 2 + 15.68 - 3/2 \lg T - \frac{5040}{T} I_{NO} \qquad (VIII)$$

E. g., using values of $I_{NO} = 9.276$ eV, $K_p = 3.60 \cdot 10^{-3}$, $T = 2500K$, $n_\mu \equiv n_{NO} = 7 \cdot 10^{16}$ particles/cm^3 (2.4 % by volume, by some other data - 2.0 %), we obtain $n_{NO^+} = n_e = 4.2 \cdot 10^9$ particles/cm^3.

Hence, it is evident that at 2500K the calculation gives a very high value of ion concentration. The true ion concentration, as it has been mentioned above, can be even somewhat higher. Such an ion concentration, as it has been discussed earlier, provides for flow of current adequate to maintain endothermic burning and to heat up the ball of lightning. Note, that for $T = 2000K$: $n_{NO} = 3 \cdot 10^{16}$ particles/cm^3 and $n_{NO^+} = n_e = 10^7$ which corresponds to the current some two orders lower as comared to the current at 2500K.

The lightning heat is transfered into environment, and the temperature inside the ball is distributed in the quasistationary way. Let's consider a simplified model of the gaseous ball ("droplet") in which the energy, q, is emitted due to the electric current passage. The value of q is equal to the Joulean heat minus the reaction endothermic heat.

For the spherically symmetric problems the stationary equation of heat conductivity at presence of a heat source with power density of Q is given in the form of

$$\frac{1}{r^2} \frac{\partial}{\partial r} \left(\lambda r^2 \frac{\partial T}{\partial r} \right) + Q = 0 \qquad (IX)$$

where: λ - heat conductivity; r - radius. Solution of equation (IX) for the ball interior has the form of

$$T = T_o + \frac{Q}{6\lambda} (r_a^2 - r^2) \qquad (X)$$

where: T_o - temperature at the ball-air boundary; r - the ball radius; T - temperature at distance r from the ball center.

Assuming that q - quantity of heat generated in the conductive medium of the lightning ball and is determined by the Joule-Lenz law

$$q = 0.24\, I_1 U t$$

where: I_1 - current (ampere); U - voltage (volt); t - time (second). Given that $\vec{i} = 4 \cdot 10^{-5}$ A/cm^2, $|\nabla \varphi| = 10^4$ V/cm, we get $Q = 9.6 \cdot 10^{-2}$ cal/cm^{3}s. Further, at $T_0 = 300$K, $\lambda = 8 \cdot 10^{-4}$ cal/cm·s·K, and $r_a = 10$cm, we get the temperature value at the ball center equal to 2300K, and the temperature 1cm away from the ball surface 680K.

To present this in a more obvious way, estimation of the air molecules diffusion time, τ, i. e. time for the air molecules diffusion into the sphere with $r_a = 10$cm at mean temperature of 2000K ($\tau = r_a^2/\text{const} \cdot D$, where D - diffusion coefficient), easily shows that the quantity of the heat absorbed in the course of the diffusion-limited reaction (1) is a small fraction (about 1 %) of the q value required for the sphere heat up. Thus, for the model being discussed, q is consumed actually for the gaseous ball heat up only.

The quantitative evaluations listed above lead us to the conclusion that under the real conditions the atmospheric volumetric charges pass through the ball lightning and maintain it. Our approach in terms of the phenomenon physics correlates with the findings discussed in some other publications[4, 26-29] considering the possiblility of the transversal "compression" of the cloud-to-ground dc current (equation IV) in the high conductivity area, and noting that the ball (sphere) positioning into a dc atmospheric field E_0 triples field intensity in the sphere. However, it should be noted that the above mentioned and other studies have not payed the proper attention to the specific role of the atmospheric volumetric charges (discharge between which in no way can be connected neither to thunderclouds nor to the ground), and have not offered a realistic physico-chemical mechanism of ion formation and maintenance of high concentrations of charges in the ball lightning. Disregard to the "chemistry of the phenomenon" has, probably, led a number of authors to overestimation of the temperatures in the ball of lightning.

As it has been widely known, quite strong horizontal atmospheric currents are able, to our estimates, to maintain the ball lightning. Those currents are especially high when volumetric charges are generated at the lightning storms, fogs, smogs, snow storms, dust storms. Those charges can maintain burning of the ball lightning same as they can induce illumination of electric bulbs.[26)]

Decrease of voltage and current in the atmosphere causes the ball lightning death (disappearance). However, at the critical discharge values (in a lightning storm) an explosion may happen or a linear lightning discharge (sparkle, arch) with an enormous amount of energy release - the ball lightning dies in the explosion. It is evident, that though the ball lightning has a relatively small stock of interior energy, it can initiate high energy discharges in the atmosphere. In principle, there is yet another possibility of a ball lightning death - a local explosion of a "chemical nature" with no discharge of a significant power - the energy of such an explosion is relatively small. It should be noted, that analysis of I, VI, X type equations shows that a stable regime of burning in the lightning can be obtained by variation of some parameters (n_{NO}, n_{NO^+}, D_i, u_i, λ, U, $\vec{i}$, r_a, etc.) in a narrow range. The bulk of those parameters are interrelated. It is clear now why the ball lightning is such a rare phenomenon with such a short lifetime.

A more detailed theory of the phenomenon can be developed with regard to the general approaches to the heat explosions and to the break down of dielectrics[30,31)] in the similar way as it has been applied to other processes.[32-34)]

The model developed by us, though doesn't account for some factors (e. g. convective transfer of matter, electric ionization degree, energy loss at irradiation, side reactions in the cold zones, etc.), is basically substantiated with reliable experimental data and is capable to explain many specific features of creation, behaviour and death of both the ball lightning and the unidentified objects.[11)] The proposed model can be easily checked experimentally. A number of concepts quoted in this study have been offered by some researchers earlier. However, the existing theoris, as a rule, have not been substantiated with the real chemistry of the lightning storm atmosphere and account for only individual aspects of the phenomenon being discussed (either only physical or desciptional chemical ones, etc.). The authors of the theories have not offered any physico-chemical mechanism of ion formation and maintenance of the required charge concentration in the ball lightning. Our model explains a range of the known facts and concepts and actually presents a detailed physicochemical model of the phenomenon.

The proposed model doesn't in principle contradict to many well known data and new findings as well as to calculations by various authors.[35-38] It is quite probable that the ball lightning phenomenon is associated with the mechanisms of the processes that have not been discussed in this paper. However, the author believes that the ball lightning is indeed a diffusional flame fed by external energy source. The ball lightning may well be considered as similar to the plasma flame created at combustion of nitrogen in industry at the nitric acid production.

References

1) Chalmers, J. A., Atmosphernoe electrichestvo, Gidromet. izd., L., 1974

2) Shishkin, N. S., Jblaka, osadki i grozovoe electrichestvo, Gidromet. izd,, 1964, 401 p.

3) Singer, S., The Nature of Ball Lightning, New York - London, Plenum Press, 1971.

4) Uman, M. A., Lightning, McGrow-Hill, New York, St. Sowis, San Francisco, Toronto, Sidney,1969 (Russian translation MIR publishers, M., 1972, 328 p.).

5) Frenkel, Ya. I., Teoria osnovnykh yavleniy atmosphernogo elevtrichestva, Izv. AN SSSR, ser. geogr. i geofiz,, 8, No. 5, 244-272, 1944

6) Nauer, H., Z. Phys., Modellversuche zum Kugelblitz, 5, No. 12, 1953, 441-450.

7) Veitsekhovskiy, B. B., Ogni Elma i svechenie na predmetakh v oblake electricheski zaryazhennykh kapel vody, DAN SSSR, 262, No. 1, 84-88, 1982.

8) Sutugin, A.,G., Petryanov, I. V., Khimicheskie prevrascheniya okislov ugleroda i azota v atmosfere, Fizika atmosfery, No. 2, 28-33, Vilnus.

9) Khrgian, A. Kh., Kuznetsov, G. I., Problema nablyudeniy i issledovaniy atmosphernogo ozona, Izd. MGU, 1981, 216 p.

10) Elanskiy, E. F., Trutce, Yu. L., Nekotorye osobennosti raspredeleniya obschego soderzhaniya ozona i dvuokisi azota v atmosfere, Izd. AN SSSR,Fizika atmosfery i okeana, 15, No. 1, 119-121, 1979.

11) Crew, E. W., Meteorological Flying Objects, Q. H. R. astr. Soc., 21, 216-219, 1980.

12) Singer, S., Ball Lightning, Naturwissenschaften, 67, 332-337, (1980).

13) Gurvich, L. V., Karachevtsev, G. V., Kondratjev, V. N., Lebedev, Yu. A., Medvedev, V. A., Potapov, V. K., Khodeev, Yu. S., Energiya razryva khimicheskikh svyazei. Potentsialy ionizatsii i srodstvo k electronu, M., Nauka,

1974, 351 p.
14) Virin, L. I., Dzhagatspanyan, V. V., Karachevtsev, G. V., Potapov, V. K., Talrose, V. L., Ionno-moleculyarnye reactsii v gasakh, M., Nauka, 1979, 548 p.
15) Zeldovitch, Ya. B., Barenblatt, T. I., Librovitch, V. B., Makhviladze, G. M., Matematicheskaya teoriya goreniya i vzryva, M., Nauka, 1980, 478 p.
16) Zeldovitch, Ya. B., Sadovnikov, P. Ya., Frank-Kamenetsky, D. A., Okisleniye azota pri gorenii, Izd. AN SSSR,M., L., 147 p.
17) Bauer, Z., Fizika planetnykh ionospher, M., MIR, 1976, 251 p.
18) Raizer, Yu. P., Osnovy sovremennoy fiziki gazorazryadnykh processov, M., Nauka, 1980, 416 p.
19) Lauton, J., Vainberg, F., Electricheskie aspecty goreniya, M., Energiya, 1976, 294 p.
20) Leng, K., Astrofizicheskie formuly, chast I, M., MIR, 326-331 p., 1978.
21) Goddard, D. R., Hughes, E. D., Ingold, C. K., J. Chem. Soc., 2539, 1950.
22) Usanovitch, M. I., Issledovaniya v oblasti teorii rastvorov i teorii kislot i osnovaniy. Izbrannye trudy. Alma-Ata, Nauka, 1970, 363 p.
23) Gladyshev, G. P., Fiziko-khimicheskoye issledovanie nekotorykh sistem, soderzhaschikh azotnuyu i sernuyu kisloty, Alma-Ata, Kaz, GU, 1959, 124 p.
24) Selezneva, I. K., Sfericheski simmetrichny opticheskiy razryad kak analog diffusionnogo goreniya v smesi goryuchikh gasov, In: Goreniye i vzryv, M., Nauka, 396-400, 1972.
25) Filippov, Yu. V., Electrosintez ozona, Vestnik MGU, No. 4, 153-186, 1959.
26) Tverskoy, P. N., Atmosphernoye electrichestvo, L., Gidromet. izd., 1949, 252 p.
27) Finkelstein, D., Rubinstein, J., Ball Lightning, Phys. Rev., 135, 2A, 390-396, 1966.
28) Andersen, W. H., Energy Source for Ball Lightning, J. Geophys. Res., 70, 6, 1291-1293, 1965.
29) Uman, M. A., Helstrom, C. W., A Theory of Ball Lightning, J. Geophys. Res., 71, 8, 1975-1984, 1966.
30) Semenov, N. N., K teorii kinetiki khimicheskikh reaktsiy, J. Rus. fiz. -khim. obschestva, 50, No. 6, 533-537, 1928.
31) Fok, V. K., K teplovoy teorii electricheskogo proboya, Tr. Leningr. fiz. -tekhn. laboratorii, 5, 52-71, 1928.
32) Velikhov, E. P., Dykhne, A. M., In; "Trudy VII Simpoz. po ioniza tsionnym yavleniyam v gazakh, Belgrad, 1967, p. 47.
33) Merzhanov, A. G., Rumanov, E. N., Teplovye processy tipa goreniya v fizike. Preprint IKHF AN SSSR, Chernogolovka, 1975, 21 p.
34) Fristrom, R. M., Vestenberg, A. A., Struktura plameni, M., Metallurgiya, 1969, 363 p.
35) Stakhanov, I. P., O fizicheskoy prirode sharovoy molnii, M., Atomizdat, 1985.
36) Smirnov, B. M., Priroda, 1987, No. 2, p. 15-26.
37) Smirnov, B. M., Problema sharovoy molnii, M., 1987.
38) Barry, James D., Ball Lightning and Bead Lightning, Extreme Forms of Atmospheric Electricity, N. Y., L., Plenum Press, 1980 (Russian translation: M., MIR, 1983, 288 p.).

NEW APPROACH TO BALL LIGHTNING

Peter H. Handel
Physics Department, University of Missouri
St. Louis, MO 63121, USA

Abstract
A new ball lightning approach based on the creation of a population inversion in some rotational energy levels of the water vapor present in a volume of air of the order of $10^2 - 10^9 m^3$ after a very strong and short electric field pulse generated by lightning is proposed, describing ball lightning as a caviton, or localized high field soliton, located somewhere in a antinode of the maser-generated standing wave. This maser-caviton approach agrees with the ball lightning phenomenology.

1. Description of The Mechanism

The rotational energy levels of the water molecule include many closely spaced pairs linked by forbidden transitions in the VHF domain. A sudden, very strong and short electric field pulse, caused by lightning, can induce a population inversion on some of these closely spaced pairs of energy levels. In the absence of very high pointed conductors, the equipotential surfaces are planes, and the population inversion can develop over volumes of many cubic kilometers. When the electric field increases to values of the order of 10^4 V/m, some of the forbidden transitions will become weakly allowed and will feed a large atmospheric maser which exceeds the critical volume. The latter is defined large enough to yield more photons per unit time through stimulated emission than the loss rate which is mainly caused by the absence of a cavity.

Due to this maser action, a standing electromagnetic wave will be present below the cloud in the vicinity of the earth, as in Kapitsa's model. If in a certain point the local plasma frequency of ions coincides with the maser frequency, a resonance process will lead to local enhancement of the electric field, to ponderomotive forces which pump out the ions from the high field region, creating a resonant cavity in the plasma, finally leading to the formation of a plasma soliton known as high-pressure caviton (Handel and Schneider, 1985). The caviton is an eigenfunction-solution of the non-linear Schroedinger equation describing the field-plasma interaction with the ponder-motive force included, physically represented by an almost empty sphere containing the resonant high electric field and ions. These cavitons were thought to be unstable with respect to collapse until recently Liedtke and Spatschek (1984) proved them to be stable for sizes exceeding a certain minimum. The calculations presented here in support of this mechanism include a calculation of the population inversion resulting from a sudden, strong and short electric field pulse as a function of its duration in units of the resonance period and an identification of the cavitonic eigenfunction and eigenfrequency. A mathematical description of the initial resonance igniting the caviton was given elsewhere (Handel and Schneider, 1985).

The caviton is thus a stable quasi-spherical configuration of trapped electromagnetic field, surrounded by plasma and moving horizontally at the level of a standing field antinode, being noticed as ball lightning. The sudden demise of the caviton fed by a large atmospheric maser in open air will lead to spiking of the maser, observed as an instantaneous large release of energy like a giant spark, and interpreted as a powerful explosion. On the other hand, the sudden demise of the caviton in closed spaces (inside a house, airplane, etc.) will not lead to noticeable energy releases or explosions, because the volume of the maser equals the volume of the closed space which is relatively small in this case, and the total energy associated with the population inversion in the water vapor contained in it is limited usually to a few hundred Joules. Finally we mention that due to the unfavorable geometry of the high-field cone above a mountaintop lightning research station, the inverted air mass will never become critical, no maser action will be possible, and ball lightning will never be observed. This explains the notorious frustration and skepticism of famous lightning researchers who never get to see ball lightning, while non-specialists in the flatlands report it frequently.

2. Calculation of Population Inversions from a Field Pulse

Let E be the amplitude of a rectangular electric field pulse lasting from t = 0 to $t = \tau$. Let E_2 and E_1 be the two closely spaced energy levels which are considered as possible upper and lower maser levels responsible for maser action at the frequency $\omega_0=(E_2-E_1)/\hbar$. Finally, let W = W(E) be the perturbation term introduced by the electric field pulse into the Hamiltonian of the individual molecules considered (H_2O for instance). W is constant in time from zero to τ, and is zero outside this short interval. We assume that W can induce transitions between the states E_2 and E_1, with nonzero matrix elements $W_{12} = W_{12}^*$.

If initially the molecules were in the state E_1 at t = 0, considering transitions to other more removed states insignificant, we determine the state at a time $0 < t < \tau$ in the usual way as a superposition of the states E_1 and E_2 with amplitudes a(t) and b(t), from Schroedinger's eq.

$$i\hbar\, da/dt = W_{11}a + W_{12}b \exp(-i\omega_0 t),$$

$$i\hbar\, db/dt = W_{12}^* a \exp(i\omega_0 t) + W_{22}b.$$

Denoting $\omega = \omega_0+(W_{22}-W_{11})/\hbar = 2\pi f$ and $r = [\omega^2/4 + |W_{12}/\hbar|^2]^{\frac{1}{2}}$,

with a(0) = 1 and b(0) = 0, we obtain the solution

$$a = \exp[-i(\omega/2 + W_{11}/\hbar)t]\,[\cos rt + (i\omega/2r)\sin rt]$$

$$b = -i(W_{12}/hr) \exp[i(\omega/2 - W_{22}/\hbar)t]\sin rt.$$

After $t = \tau$ the amplitudes a and b will no longer vary, but will retain the value they will have reached at $t = \tau$. The resulting population ratio R of the two levels after the field pulse is therefore

$$R = |2W_{12} \sin r\tau|^2/[(hf)^2 + |2W_{12} \cos r\tau|^2] .$$

We see that for $r\tau = (2n+1)\pi/2$ (i. e., perhaps in 10% of the cases), we obtain a population inversion ($R > 1$), provided the perturbation satisfies the criterion $2W_{12} \gg hf$. The matrix element must exceed $E_2 - E_1$. This is very likely if E_1 and E_2 are among the closely spaced pairs of rotational energy levels of the water molecule, and if the field amplitude is large with abrupt onset and termination.

3. Cavitonic Eigenfunction and Resonance Frequency

To analyze the developed caviton, we start from following three eqs.

$$\partial x/\partial t + \nabla\cdot[(n+n')\vec{v}] = 0, \tag{1}$$

$$(n+n')\ \partial\vec{v}/\partial t + (n+n')\ e\vec{E}/m + (T/m)\nabla x = 0, \tag{2}$$

$$\nabla\cdot\vec{E} = -4\pi ex. \tag{3}$$

Here we have divided the electron concentration in three components: n is the average electron concentration, constant in space and time, balanced by the average ionic concentration; n' is a very slow fluctuation, neutralized by an identical fluctuation in the ionic concentration, also including a large time-independent part defining the caviton; the third part x is the fast fluctuation which can not be followed by the ions. We have considered both x and the average electron velocity $\vec{v}$ small, and neglected terms quadratic in them. The first equation is the equation of continuity for the fast (jitter) motion, the second is the equation of motion expressing the momentum balance, with the electron temperature T in energy units. Eq. (3) is Poisson's equation. We neglect the time-derivatives of n'. m is the mass of the electron. Taking the divergence of Eq. (2), subtracting the time-derivative of Eq. (1), and eliminating x with Eq. (3), we obtain

$$\partial^2 E/\partial t^2 + 4\pi e^2(n+n')\vec{E}/m - T\nabla(\nabla\cdot\vec{E}) = 0. \tag{4}$$

The fast part of the variation in E(t) can be expressed through a time-dependent exponential

$$E(t) = E_0(t)\ \exp(-i\omega_p t) + \text{c.c.}, \tag{5}$$

where E_0 is a slowly varying amplitude with a negligible second derivative, $\omega_p = 4\pi\ ne^2/m$, and c. c. is the complex conjugate of the preceding term. The ponderomotive force of the electric field E(t) on the plasma of electrons and ions can be derived from an average potential energy ($\omega \simeq \omega_p$ is the actual "fast" frequency)

$$U = |eE_0|^2/4m\omega^2 = e^2 \langle E^2 \rangle /2m\omega^2 , \tag{6}$$

where $\langle\ \rangle$ is a time average. Denoting by T_i and T_e the ionic and electronic temperatures in energy units, we expect a Boltzmann distribution of the plasma density

$$n + n' = n_0 \exp[-U/(T_i + T_e)] . \tag{7}$$

Introducing the dimensionless field amplitude $\vec{A} = e\vec{E}_0/2\omega[m(T_i + T_e)]^{\frac{1}{2}}$, we obtain

$$2i\omega_p \partial\vec{A}/\partial t + \omega_p^2 \vec{A}[\exp(-|A|^2) - 1] - 1_D^2\ \omega_p^2 \nabla(\nabla\cdot\vec{A}) = 0. \tag{8}$$

Introducing the dimensionless variables $\tau = t\omega_p/2$ and $s = r/1_D$, where $1_D = (T_e/4\pi ne^2)^{\frac{1}{2}}$ is the Debye length, we get the fundamental nonlinear "Schroedinger" equation (Zaharov, 1972)

$$i\partial\vec{A}/\partial\tau + \nabla_s(\nabla_s\cdot\vec{A}) + [1 - \exp(-|A|^2)]\vec{A} = 0, \tag{9}$$

which describes both the temporal evolution and the stationary states of the solitons (field structures trapped in the plasma) in terms of the normalized field amplitude $\vec{A}$. For stationary spherically symmetric solutions $\vec{A} = \vec{X}(s)\exp(-i\Omega t)$, one obtains ($\vec{X} = X\ \vec{s}/s$)

$$s^{-2}(\partial/\partial s)[s^2(\partial X/\partial s)] - 2X/s^2 + [1 - \exp(-|X|^2)]\ X = \Omega X \tag{10}$$

with the boundary conditions X = 0 for $s = \infty$ and X bounded for s = 0. Like the stationary Schroedinger eq., this is an <u>eigenvalue equation</u> for Ω. This equation admits solutions which we will call cavitons and which are similar to the bound states of quantum mechanics: fundamental (n = 0) and excited (n = 1, 2, ...) modes. Two numerically calculated solutions of Eq. (10) by J.

Zitkova Wilcox and T. J. Wilcox (1975) for n = 0 and n = 1 are shown in the figure below. The stability of these solitons when they exceed a certain critical size was proven by Liedtke and Spatschek (1984).

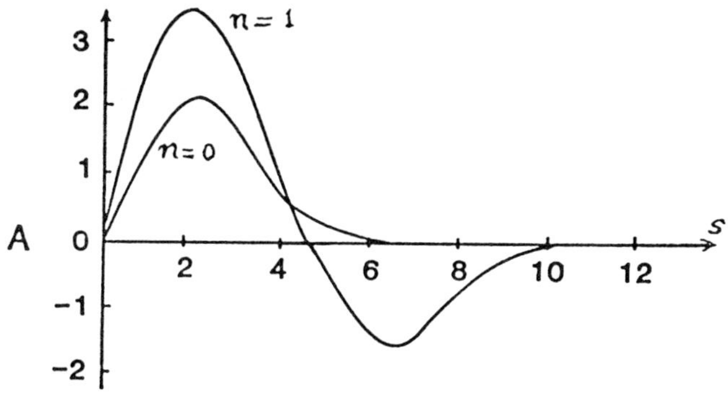

References

Handel, P. H. (1985), Polarization catastrophe theory of cloud electricity. J. Geophys, Res. 90, 5857-5863.

Handel, P. H. and Schneider, R. T. (1985). Nature of Resonances Leading to High-Pressure Cavitons, Fusion Technology 7, 320-324

Liedtke, H. and Spatschek, A. (1984). Stability of Plasma Cavitons, Proc. Int. Conf. on Plasma Physics and Fusion, Lausanne, Switzerland.

Zaharov, V. E. (1972), Zh. Eksp. Teor. Fiz. 62, 1745

Zitkova Wilcox, J., Wilcox, T. J. (1975), Phys. Rev. Lett. 34, 1160.

A Calculation of Electric Field of Ball Lightning

T. Neda[1), H. Ofuruton[2) and Y. H. Ohtsuki[1)

[1)Department of Physics, Waseda University,
Ohkubo 3-4-1, Shinjuku-ku, Tokyo 160, Japan

[2)Tokyo Metropolitan Collage of Aeronautical
Engineering, Minamisenju 8-53-1,
Arakawa-ku, Tokyo 116, Japan

The authors calculated the relative magnitude of the electric field inside and outside the ball lightning (BL) using Mukharev's aerosol model which has many advantages. Some shapes of BLs are explained by the calculation result. And the calculated life time agrees with the observed one very well.

I. INTRODUCTION

In 1985 Mukharev proposed the aerosol model of the BL. In this model the BL is thought to be an atmospheric phenomenon with the internal source of energy. It is assumed to be created in a charged aerosol medium by an exponentially damped electromagnetic field. The electromagnetic field excites a luminous, spherical and electrode-free coronal discharge.

Usually about 30 % of aerosol particles in the atmosphere have an positive elementary electric charge respectively. And about 30 % have negative elementary electric charge. The rest are neutral. An experiment showed the property that the cluster of charged aerosol particles is spherical.

When lightning discharge passes through the cluster the aerosol particles are given kinetic energy. This pumped energy is the source of the BL activity.

In the spherical medium the electromagnetic field is represented as the sum of fields of spherical harmonics. When some of the modes are strongly excited the spherical corona discharge can be excited. One of the condition for the strong excites of some modes is the existence of spherical area in which the dielectric constant differs from that of environment. In such a case the amplitude of the field becomes maximum at the boundary spherical surface.

The energy source of the BL is the kinetic energy of aerosol particles given by the lightning discharge. The energy decreases by the collision between aerosol particles and air molecules and by the radiation. Then the electromagnetic field damps exponentially. After the amplitude damps to the limit of corona discharge 10^6 V/m, the discharge stops and the BL fades without sound and explosion.

The aerosol model has many advantages. BLs are often observed with thunderstorm. This agrees with this model because the lightning discharges are thought to excite the BL. Most of observations of the BL are under low temperature and under the atmospheric pressure. This agrees with the property of the corona discharge. Sometimes the BL with the odor of O_3 or NO_x is observed. This odor can be also thought to be caused by corona discharge. For most of observations the BL fades without sound and explosion. This is in agreement with this model too. In a few observations, after the BL faded, there remained white smoke. This is assumed to be the cloud of aerosol clusters.

II. THEORY

The amplitude of the electromagnetic field of the BL can be calculated using Mukharev's aerosol model.

The authors imagine a sphere with radius a in the atmosphere. Inside the sphere there are charged aerosol particles with density N, mass m_a and charge $\pm q$. The half has positive charge +q and the rest half has negative charge −q respectively. As a whole the sphere is neutral. Thus this aerosol medium can be regarded as the plasma medium. Outside the sphere is imagined as the perfect dielectric. Both inside and outside the sphere the magnetic permeability is regarded as the vacuum value μ_0. And the electric and magnetic fields inside the sphere are written as $\mathbb{E}^{(1)}(\mathbb{r}, t)$ and $\mathbb{H}^{(1)}(\mathbb{r}, t)$. In the case of the outside $\mathbb{E}^{(2)}(\mathbb{r}, t)$ and $\mathbb{H}^{(2)}(\mathbb{r}, t)$.

The complex dielectric constant ϵ_1 inside the sphere can be calculated through the equation of motion

$$m_a \frac{d\mathbb{v}}{dt} = \pm q \mathbb{E}^{(1)}(\mathbb{r}, t) - m_a \nu \mathbb{v} \tag{1}$$

where $\mathbb{v}$ is the velocity of the aerosol particles and ν is the collision frequency between the aerosol particles and the air molecules. The time dependence of $\mathbb{E}^{(1)}$ can be assumed to be $\exp(-i\omega t)$. Then the complex dielectric constant ϵ_1 is

$$\epsilon_1 = \epsilon_0 \left\{ 1 + \frac{\omega_{pa}^2}{\nu^2 + \omega^2} \left(i\frac{\nu}{\omega} - 1 \right) \right\} \tag{2}$$

where ω_{pa} is the ion plasma frequency $\sqrt{Nq^2/m_a \epsilon_0}$.

The vector Helmholtz equations are derived from Maxwell's equations and so.

$$(\nabla^2+k_1^2)\,\mathbb{E}^{(1)}(\mathbb{r},t)=0 \tag{3}$$

$$(\nabla^2+k_1^2)\,\mathbb{H}^{(1)}(\mathbb{r},t)=0 \tag{4}$$

$$k_1^2=\mu_0\epsilon_1\omega^2 \tag{5}$$

$$(\nabla^2+k_2^2)\,\mathbb{E}^{(2)}(\mathbb{r},t)=0 \tag{6}$$

$$(\nabla^2+k_2^2)\,\mathbb{H}^{(2)}(\mathbb{r},t)=0 \tag{7}$$

$$k_2^2=\mu_0\epsilon_0\omega^2 \tag{8}$$

The authors solved these equations with spherical coordinates and chose the solutions which are finite both at r=0 and at $r=\infty$. Then these solutions can be divided into two groups. One is transverse electric modes (TE mode) and the other is the transverse magnetic modes (TM mode). Only TM modes were considered because of advantages of the approximation. The solution of TM modes is

$$(\mathbb{E}^{(1)}_{nm})_r = -n(n+1)\frac{j_n(k_1r)}{k_1r}Y_n^{m(1)}(\theta,\phi)\,e^{-i\omega t} \tag{9}$$

$$(\mathbb{E}^{(1)}_{nm})_\theta = -\left\{(n+1)\frac{j_n(k_1r)}{k_1r}-j_{n+1}(k_1r)\right\}\frac{\partial Y_n^{m(1)}(\theta,\phi)}{\partial\theta}\,e^{-i\omega t} \tag{10}$$

$$(\mathbb{E}^{(1)}_{nm})_\phi = -\left\{(n+1)\frac{j_n(k_1r)}{k_1r}-j_{n+1}(k_1r)\right\}\frac{1}{\sin\theta}\frac{\partial Y_n^{m(1)}(\theta,\phi)}{\partial\phi}\,e^{-i\omega t} \tag{11}$$

$$(\mathbb{E}^{(2)}_{nm})_r = -n(n+1)\frac{h_n^{(1)}(k_2r)}{k_2r}Y_n^{m(2)}(\theta,\phi)\,e^{-i\omega t} \tag{12}$$

$$(\mathbb{E}^{(2)}_{nm})_\theta = -\left\{(n+1)\frac{h_n^{(1)}(k_2r)}{k_2r}-h_{n+1}^{(1)}(k_2r)\right\}\frac{\partial Y_n^{m(2)}(\theta,\phi)}{\partial\theta}\,e^{-i\omega t} \tag{13}$$

$$(\mathbb{E}^{(2)}_{nm})_\phi = -\left\{(n+1)\frac{h_n^{(1)}(k_2r)}{k_2r}-h_{n+1}^{(1)}(k_2r)\right\}\frac{1}{\sin\theta}\frac{\partial Y_n^{m(2)}(\theta,\phi)}{\partial\phi}\,e^{-i\omega t} \tag{14}$$

where j_n and $h_n^{(1)}$ are the spherical Bessel functions and Y_n^m is the spherical harmonics

$$Y_n^{m(1)}(\theta,\phi) = (A_{nm}^{(1)}\cos m\phi + B_{nm}^{(1)}\sin m\phi)\,P_n^m(\cos\theta) \tag{15}$$

$$Y_n^{m(2)}(\theta,\phi) = (A_{nm}^{(2)}\cos m\phi + B_{nm}^{(2)}\sin m\phi)\,P_n^m(\cos\theta) \tag{16}$$

The boundary conditions should be considered. At the boundary where r=a the tangential components of the electric field and the magnetic field should be continuous. The radial components of the electric flux density and the magnetic flux density should be also continuous.

$$\left\{(n+1)\frac{j_n(k_1a)}{k_1a} - j_{n+1}(k_1a)\right\}A_{nm}^{(1)} = \left\{(n+1)\frac{h_n^{(1)}(k_2a)}{k_2a} - h_{n+1}^{(1)}(k_2a)\right\}A_{nm}^{(2)} \tag{17}$$

$$k_1 j_n(k_1a)\,A_{nm}^{(1)} = k_2 h_n^{(1)}(k_2a)\,A_{nm}^{(2)} \tag{18}$$

$$\left\{(n+1)\frac{j_n(k_1a)}{k_1a} - j_{n+1}(k_1a)\right\}B_{nm}^{(1)} = \left\{(n+1)\frac{h_n^{(1)}(k_2a)}{k_2a} - h_{n+1}^{(1)}(k_2a)\right\}B_{nm}^{(2)} \tag{19}$$

$$k_1 j_n(k_1a)\,B_{nm}^{(1)} = k_2 h_n^{(1)}(k_2a)\,B_{nm}^{(2)} \tag{20}$$

From these conditions the equation

$$\frac{1}{k_1}\left\{(n+1)\frac{1}{k_1a} - \frac{j_{n+1}(k_1a)}{j_n(k_1a)}\right\} = \frac{1}{k_2}\left\{(n+1)\frac{1}{k_2a} - \frac{h_{n+1}^{(1)}(k_2a)}{h_n^{(1)}(k_2a)}\right\} \tag{21}$$

is derived. ω, k_1 and k_2 are determined through equations (5), (8) and (21). But undefined coefficients $A_{nm}^{(1)}$, $B_{nm}^{(1)}$, $A_{nm}^{(2)}$ and $B_{nm}^{(2)}$ are left undetermined. Only their ratio are determined. To determine their values the initial condition is indispensable. Up to now the initial condition has not been obtained observationally. Then the relative magnitude of the electric field is considered here.

The parameter values are given as follows. The radius of the sphere $a=1.0\times10^{-1}$ m which is almost the mean value of the BL. The mass of the aerosol particle $m_a=8.7\times10^{-14}$ kg which is the value for the quartz dust. The charge of the aerosol particle $q=1.6\times10^{-15}$ C which is 10^4 times as large as that of the air in cities. The density of the aerosol $N=2.3\times10^9$ m^{-3} which is 10^3 times as large as that of the air in cities and as large as that of the soot.

Using these values the ion plasma frequency $\omega_{pa}=87.4$ s^{-1}. The aerosol particles can be regarded as spheres with radius r_a. We assume $r_a=2.0\times10^{-6}$ m which is the value for the particle of the incense. And air molecules can be regarded as the rest fluid with the viscosity $\eta=1.7\times10^{-5}$ Nsm^{-2}. Then ν is calculated as $\nu = 6\pi\eta r_a/m_a = 7.4\times10^3$ s^{-1}.

Using these values the authors calculate the relative magnitude of the electric field. The equation (21) is the transcendental equation with complex coefficients. Then approximations are necessary. In the case of $|\omega|<<10^8 s^{-1}$, $|k_1a|<<1$ and $|k_2a|<<1$ are valid, we can use the approximations

$$j_n(k_1r) \approx \frac{(k_1r)^n}{(2n+1)!!} \tag{22}$$

$$h_n(k_2r) \approx -i(2n+1)!!\,(k_2r)^{-n-1} \tag{23}$$

Then equation (21) is approximated to

$$\frac{n+1}{(k_1a)^2}+\frac{n}{(k_2a)^2}=\frac{1}{2n+3} \tag{24}$$

Solving these equations, following approximate solutions are obtained.

$$\omega \approx -i\frac{n}{2n+1}\frac{\omega_{pa}^2}{\nu} \tag{25}$$

$$k_1 \approx \pm i\sqrt{\frac{n+1}{n}}\,\frac{\omega}{c_0} = \pm\frac{\sqrt{n(n+1)}}{2n+1}\,\frac{\omega_{pa}^2}{c_0\nu} \tag{26}$$

$$k_2 \approx -i\frac{\omega}{c_0} = i\,\frac{n}{2n+1}\,\frac{\omega_{pa}^2}{c_0\nu} \tag{27}$$

Under this approximation the electric field is expressed as follows:

$$(\mathbb{E}_{nm}^{(1)})_r \approx -n\frac{n+1}{(2n+1)!!}(k_1r)^{n-1}\,Y_n^{m(1)}(\theta,\phi)\,e^{-i\omega t} \tag{28}$$

$$(\mathbb{E}_{nm}^{(1)})_\theta \approx -\frac{n+1}{(2n+1)!!}(k_1r)^{n-1}\,\frac{\partial Y_n^{m(1)}}{\partial\theta}(\theta,\phi)\,e^{-i\omega t} \tag{29}$$

$$(\mathbb{E}_{nm}^{(1)})_\phi \approx -\frac{n+1}{(2n+1)!!}(k_1r)^{n-1}\,\frac{1}{\sin\theta}\,\frac{\partial Y_n^{m(1)}(\theta,\phi)}{\partial\phi}\,e^{-i\omega t} \tag{30}$$

$$(\mathbb{E}_{nm}^{(2)})_r \approx in(n+1)(2n-1)!!\,(k_2r)^{-n-2}\,Y_n^{m(2)}(\theta,\phi)\,e^{-i\omega t} \tag{31}$$

$$(\mathbb{E}_{nm}^{(2)})_\theta \approx -in(2n-1)!!\,(k_2r)^{-n-2}\,\frac{\partial Y_n^{m(2)}}{\partial\theta}(\theta,\phi)\,e^{-i\omega t} \tag{32}$$

$$(\mathbb{E}_{nm}^{(2)})_\phi \approx -in(2n-1)!!\,(k_2r)^{-n-2}\,\frac{1}{\sin\theta}\,\frac{\partial Y_n^{m(2)}(\theta,\phi)}{\partial\phi}\,e^{-i\omega t} \tag{33}$$

III. RESULTS AND DISCUSSION

Using expressions from (25) to (33) the authors calculate the relative magnitude of the electric field of the BL E. And they draw figures (Fig. 1–Fig. 6) representing the magnitude E for the plane where $\phi=0$,π at t=0.

The circular line shows the aerosol sphere with radius a. The magnitude E has its maximum value at r=a. Inside the sphere $E\propto r^{n-1}$ then E is uniform for the 1-0 mode (n=1, m=0) and the 1-1 mode while E decreases as r decreases for the other modes. Outside the sphere $E\propto r^{-n-2}$ then E decreases as r decreases. In the figures the area where E is larger is represented the darker black area. To show the distribution of E clearly the white areas are put between black areas.

In fact the field consists of the sum of some modes. But as the special case the case that only one mode is excited is imagined. Suppose that only 1−0 mode or 1−1 mode is excited, the observed BL is assumed to be the luminous sphere or ellipsoid. This is the shape of the ordinary BL. Suppose that only 2−0 mode or 2−1 mode is excited, the observed BL is assumed to be the shape of luminous torus or sphere with darker center. These are almost 5 % of the reports of the BL. Suppose that only 2−2 mode is excited the observed BL is the luminous sphere with a darker line in its center. There are a few reports of the BL of this shape.

The authors calculate the life time of the BL. The frequency ω of the field has the value of -5.2×10^{-1} s^{-1}. Then the life time which is the reciprocal of $i\omega$ is 1.9 s. This agrees with the observed one.

IV. CONCLUSION

The calculation based on the present aerosol model can explain various shapes of the BL which correspond to various types of observation reports, including quite rare ones. And the calculated life time corresponds well to observed one.

V. REFERENCE

1) L.A.Mukharev, Sov.J.Commun.Technol.& Electron 30 (9) 77 (1985)

2) J.A.Stratton, 'Electromagnetic Theory' (1941) McGraw-Hill

3) K. Takahashi, 'Fundamentals of Aerosol Technology' (in Japanese) (1982), Yohkendo (Japan)

Figure Captions

Fig.1

The relative magnitude of the 1-0 mode at the plane where $\phi=0,\pi$ at $t=0$.

Fig.2

The relative magnitude of the 1-1 mode at the plane where $\phi=0,\pi$ at $t=0$.

Fig.3

The relative magnitude of the 2-0 mode at the plane where $\phi=0,\pi$ at $t=0$.

Fig.4

The relative magnitude of the 2-1 mode at the plane where $\phi=0,\pi$ at $t=0$.

Fig.5

The relative magnitude of the 2-2 mode at the plane where $\phi=0,\pi$ at $t=0$.

Fig.6

The relative magnitude of the 3-1 mode at the plane where $\phi=0,\pi$ at $t=0$.

Fig.1

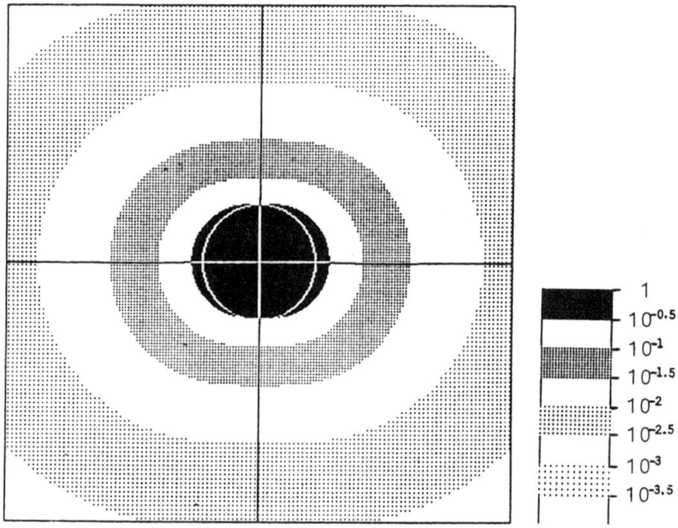

Fig.2

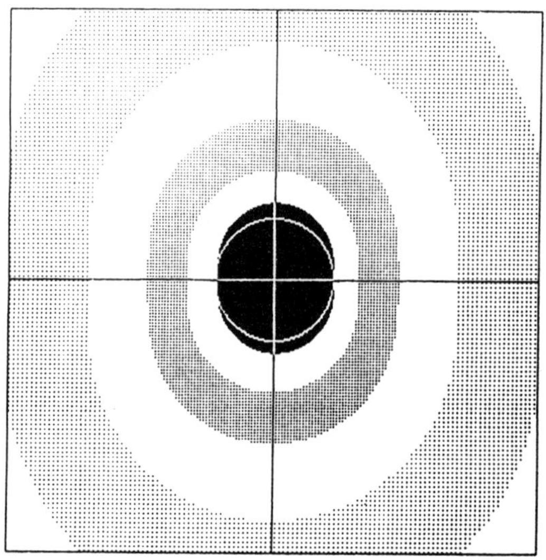

Fig.3

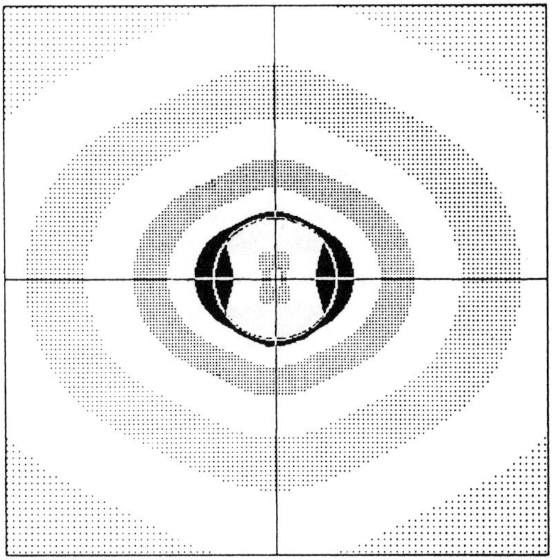

Fig.4

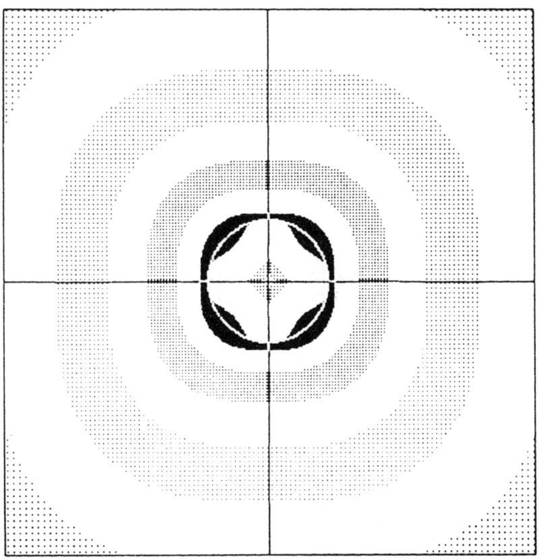

Fig.5

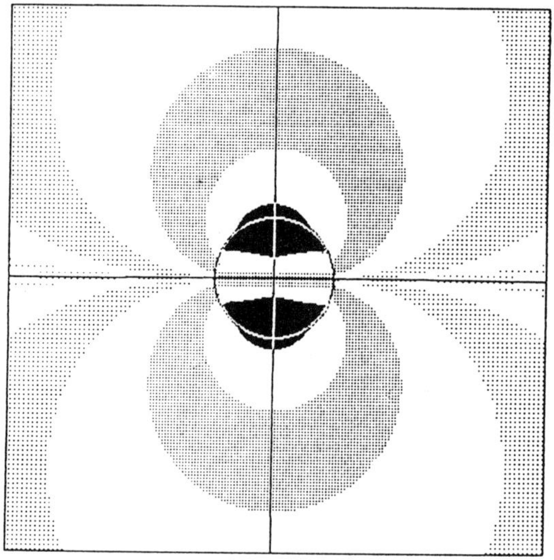

Fig.6

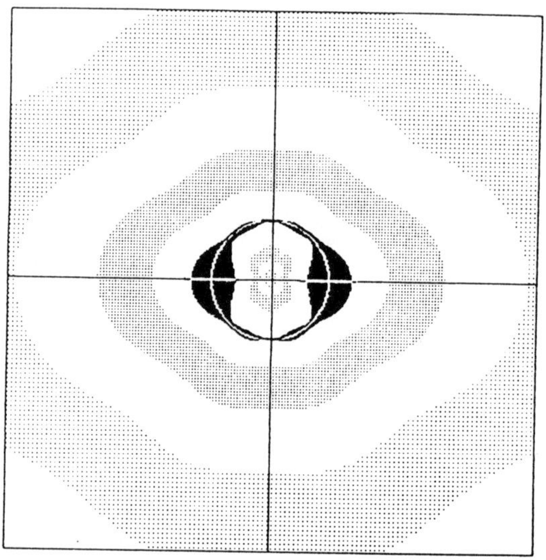

THE PHYSICAL EXPLANATION TO THE UFO OVER XINJIANG, NORTHERN WEST CHINA

Zou Yousuo
Institute of Atmospheric Physics
Academia Sinica, Beijing

Abstract
On the 18th, March, 1988, after sunset, a strange fire balls were observed in the fine sky at 11000m over Qijiaojing area in Hami in the Xinjiang Uygrur Autonomous Region of northern west China. In this article, details of the witnesses are described and compared with the results of laboratory fluid experiments as well as computer numerical studies. It is concluded that the unidentified flying object (UFO) might be the plasma vortex or plasma soliton.

1. Introduction

In recent years, witness reports of fire balls that happened in fine weather are constantly emerging in China. Up to now, we have collected 3000 witness reports in China. These fine weather fire balls displayed many unintelligible properties, therefore, people call these fire balls as UFO (Unidentified Flying Object). Scientists in China have already paid attention to these kind of fire balls in fine weather. Based on the nature of a great number of UFO witness reports and principles of plasma physics, we put forward a concept of rotating plasma to explain ball lightning, so-called fire ball, and UFO in atmosphere (Zou Yousuo, 1984). As for the relationship between ball lightning and UFO, Dr. S. Singer (1977), a famous expert of ball lightning, has contributed some discussion about it.

As an example, this article describes one case of UFO phenomenon observed by a passenger - airplane of Civil Aviation Administration of China (CAAC), and trys to give a possibly physical explanation.

2. The Details of the UFO Witness

On 20th, March, 1988, all Chinese major newspapers reported a UFO news observed on 18th March, 1988 by pilots and passengers in the airplane Tu154-2606 of Xinjiang Aviation Company of CAAC. After hearing that, the researchers at the Institute of Atmospheric Physics of Chinese Academy of Sciences immediately began to investigate the case. The following details are from Xinjiang Aviation Company.

It was at 21:35 (Beijing time) on March 18, as the No. 2606 airliner from Beijing to Urumqi was approaching Qijiaojing area in Hami in the Xinjing Uygrur Autonomous Region of northern-west China. Suddenly, pilots and passengers discovered a strange fire ball shaped like a basketball beaming a strong white ray like a searchlight, in the direction of 310°, right-hand, before and a little above the airplane. The plane was flying from east to west against current wind at 11000m high. See Fig. 1.

The UFO flying toward airplane, was gradually seen clear. The pilots opened their signal light to contact the flying object but it did not make any responses. At this moment all the meters and instruments in the plane were going well and communications between the plane and the ground facilities were not abnormal. 3 minutes later, the flying object changed its flying direction and went to the north. Meanwhile it split into two — a smaller ball above and a bean-shaped one below. Passengers saw one of the objects was spinning at high speed and was surrounded by a green ring of light. Then the two bodies were coalesced gradually into one and the surrounding green light ring remained. Then the flying object became smaller and smaller. At 21:48 (Beijing time), it disappeared in the light of stars.

At that time, the local meteorological conditions at the airline (11000m high) were as follows. The sky was clear and visibility was good with a little cloud and weak light of stars in distance. It was after sunset and dark. The terrains are Tianshan Mountains and the Gobi Desert. The lights of Hami city on the ground could be seen (Fig. 1). Wind: 100km/h of speed and $270\text{-}280^{\circ}$ of direction at 11000m height measured by the airplane. Temperature: $-50^{\circ}C$ at 11000m in height. The flying data of the airplane: 800km/h, direction 290° from the east to the west flying against current wind. Air flow field: see Fig. 2. In Fig. 2a, we can see the large scale environment of air flow field at about 12500m over Xinjiang. Fig. 2b describes air flow field at about surface ground. All meteorological data in Fig. 2a and Fig. 2b were from The Central Meteorological Observatory of China. Comparing Fig. 2b and Fig. 2a, we can see the wing shear between 1000hpa and 200hpa. Wind direction over Qijiaojing changed from about 315° to 255° while wind velocity changed from 10m/s to about 35m/s. However, the exactly local wind data near or above the UFO was not measured by local meteorological stations. We can just try to describe the local air flow field near the UFO as in Fig. 2c.according to the large scale wind field in Fig. 2a and Fig. 2b, local topography of mountains shown in Fig. 2c as well as wind data at 11000m height measured by the airplane Tu 154-2606 on 18th March, 1988.

3. Possiblly Physical Explanation

a) Peter Freymuth's fluid experiments

Freymuth (1966) published his fluid-smoke experiments for dealing with the growth of small disturbances in a shear laminar boundary layer and for the transition from laminar flow to turbulence. In his test chamber, the air fluid velocity distribution is the antisymmetric hyperbolic-tangent profile:

$$U(y) = 0.5\ (1 + \tanh 0.5y) \qquad (1)$$

which was taken as the undisturbed basic flow. And, the artificial harmonic (sine) disturbances in the shear layer of the jet were excited by a sound loudspeaker. To make photographs clear, the smoke were introduced into the shear flow by a smoke pipe. Fig. 3 is the photograph of his experiment results, showing the vortices formed and slipped into two at first, then coalesced into one again. Finally, instabilities developed, the vortices breaking down and became turbulence.

This results of experiment was well theoretically explained by R. E. Kelly (1967). Kelly consider stream function $\psi(x, y, t)$,

$$u = \frac{\partial \psi}{\partial y} \qquad\qquad v = -\frac{\partial \psi}{\partial x} \tag{2}$$

and assumed ψ consisting of basic flow ψ_0 and perturbance ψ_1:

$$\psi(x, y, t) = \psi_0(x, y, t) + \epsilon\psi_1(x, y, t) + O(\epsilon^2) \tag{3}$$

then, he obtained:

$$\left[\frac{\partial}{\partial t} + \frac{\partial \psi_0}{\partial y}\frac{\partial}{\partial x} - \frac{\partial \psi_0}{\partial x}\frac{\partial}{\partial y}\right]\nabla^2\psi_1 + \left(\frac{\partial}{\partial x}\nabla^2\psi_0\right)\frac{\partial \psi_1}{\partial y} - \left(\frac{\partial}{\partial y}\nabla^2\psi_0\right)\frac{\partial \psi_1}{\partial x} = 0 \tag{4}$$

from

$$\left(\frac{\partial}{\partial t} + \frac{\partial \psi}{\partial y}\frac{\partial}{\partial x} - \frac{\partial \psi}{\partial x}\frac{\partial}{\partial y}\right)\nabla^2\psi = 0 \tag{5}$$

and boundary conditions for (4): $\frac{\partial \psi_1}{\partial x} \to 0$ as $y \to \pm\infty$, $\frac{\partial \psi_1}{\partial y} \to 0$, as $y \to \pm\infty$ (5).

b) Solitons in vortices or in rapidly rotating fluid

Gibbon, J. D. et al. (1979) gave out the sine-Gordon soliton equation from the vorticity equation for a rapidly rotating fluid of two shear layers. He considered the following potential vorticity equation for each layer:

$$\left(\frac{\partial}{\partial t} + \frac{\partial \psi_1}{\partial x}\frac{\partial}{\partial y} - \frac{\partial \psi_1}{\partial y}\frac{\partial}{\partial x}\right)\left[\nabla^2\psi_1 + F(\psi_2 - \psi_1) + \beta y\right] = 0 \tag{6}$$

$$\left(\frac{\partial}{\partial t} + \frac{\partial \psi_2}{\partial x}\frac{\partial}{\partial y} - \frac{\partial \psi_2}{\partial y}\frac{\partial}{\partial x}\right)\left[\nabla^2\psi_2 + F(\psi_1 - \psi_2) + \beta y\right] = 0 \tag{7}$$

where ψ_1, ψ_2 are geostrophic stream function for each of two fluid layers respectively. F is froude number depends on the difference between the two fluid densities ($F = \Omega^2 R/g$ = centrifugal force/gravity). From (7) and (6), Gibbon et al. finally obtained the sine-Gordon equation

$$\psi_{\xi\eta} = \pm \sin\psi \tag{8}$$

or

$$\phi_{tt} - (\phi_{xx} + \phi_{yy}) = \pm \sin\phi \tag{9}$$

It is well known that the sine-Gordon equation well describes soliton movement. This equation was got both in plasma and fluid, there are a lot of work done on plasma solitons by using sine-Gordon equation. Here let us see the numerical study of sine-Gordon solitons finished by P. L. Christiansen and P. S. Lomdahl (1981). They used:

$$\begin{aligned} &(\phi_{xx} + \phi_{yy}) - \phi_{tt} = \sin\phi \\ &- x_0 < x < x_0, \; - y_0 < y < y_0, \quad t \geq 0 \end{aligned} \tag{10}$$

and initial conditions:

$$\begin{aligned} &\phi(x, y, t)|_{t=0} = 4 \tan^{-1}\exp(3 - \sqrt{x^2+y^2}) \\ &\phi_t(x, y, t)|_{t=0} = 0 \end{aligned} \tag{11}$$

The computation results of theirs are shown in Fig. 4. Concerning the ring soliton's slipping and coalescing, Christiansen and Lomdahl's computing results are shown in Fig. 5.

V. G. Makhankov (1979) have studied two soliton's interaction in detail with computer numerical study. His computation result for bound state of two solitons is shown in Fig. 6.

c) The production of plasma in atmosphere

How does the initial plasma for forming solitons in atmosphere come from? It is a difficult problem to answer. But the green fire-ring around the UFO over Xinjiang prompts us that the UFO-soliton plasma may come from high frequency magnetic discharge, which is because

$$- \frac{1}{c} \frac{\partial \vec{B}}{\partial t} = \nabla \times \vec{E} \tag{12}$$

The rings of E excited by HF magnetic field will heat air and turn the air into plasma. Let

$$\vec{E} = \vec{E}_0 \, e^{i\omega t} \tag{13}$$

The power from $\vec{E}$ is turned into the air within one period T of oscillation:

$$\bar{P} = \frac{1}{T} \int_0^T \sigma |\vec{E}|^2 \, dt = \frac{ne^2}{m\nu} \cdot \frac{1}{2} E_0^2 \cdot \frac{\nu^2}{\nu^2+\omega^2} \tag{14}$$

where ω is the frequency of annulus $\vec{E}$.

Another possible way for the soliton plasma may come from the earth surface and then rise to the air according to many witness reports in China.

4. Discussion

Plasma solitons have been experimented and studied for many years already in the world (Schamel and Ichikawa, 1979). As we know, if the density of plasma is not too large, plasma temperature is not too high and the electric and magnetic field are slight. We may treat the plasma as a gas or fluid. So, on the basis of the discussion above, it can be concluded that the UFO over Xinjiang may be a plasma vortex or plasma soliton. However, there is still many problems unclear, such as the initial plasma come from, etc. This article just attempt to take an example to demonstrate the fire balls that may happen in clear sky which should be noted and studied further.

The author is much graterul to Dr. S. Singer, who have given me a great deal of helps with my studies of fire balls and atmospheric electricity and encouraged me to study the connection between solitons and ball lightning.

Reference

Y. S. Zou(1984), Ball lightning earthquake fire ball and UFO — Rotating plasma in atmosphere, reported at China National Symposium on UFO, Aug. 1986: and at Symposium on Atmospheric Physics of Chinese Meteorological Society, May 1988.

S. Singer (1977), Ball lightning, Lightning edited by R. H. Golde, Academic Press (London), pp. 409-436.

P. Freymuth (1966), On transition in a separated laminar boundary layer, J. Fluid Mech. Vol. 25, Part 4, pp. 683-704.

R. E. Kelly (1967), On the stability of an inviscid shear layer which is periodic in space and time. J. Fluid Mech. Vol. 27 Part 4, pp. 657-689.

Hans Schamel (1979), Role of trapped particles and waves in plasma solitons — Theory and Application, Physica Scripta, Vol. 20, pp. 306-316.

Y. H. Ichikawa (1979), Topics on solitons in plasma, Physica Scripta, Vol. 20, pp. 296-305.

J. D. Gobbon et al. (1979), The sine-Gordon equation as a model for a rapidly rotating baroclinic fluid, Physica Scripta, Vol. 20, pp. 402-408.

P. L. Christiansen and P. S. Lomdahl (1981), Numerical study of 2+1 dimensional sine-Gordon solitons, Physica 2D, pp. 482-494.

V. G. Makhankov, G. Kummer and A. B. Shvachka (1979), Many dimensional U(1) solitons, their interaction, resonances and bound states, Physica Scripta, Vol. 20, pp. 454-461.

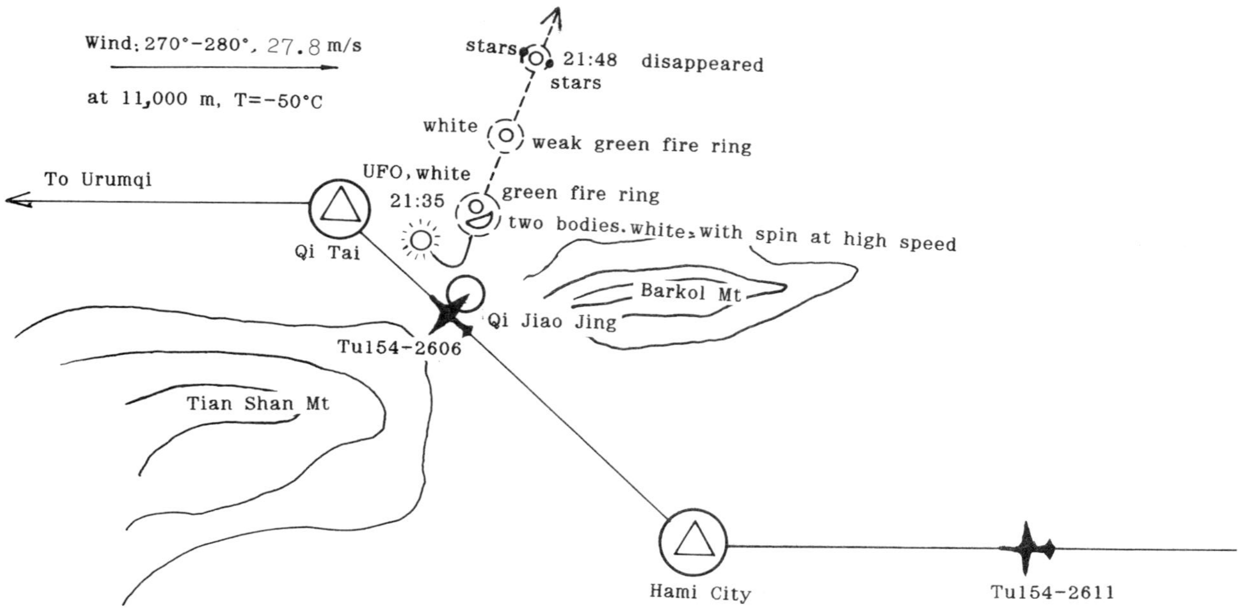

Fig.1. Sketch for the UFO on March, 18, 1988, at 11,000 m over Xinjiang, China.

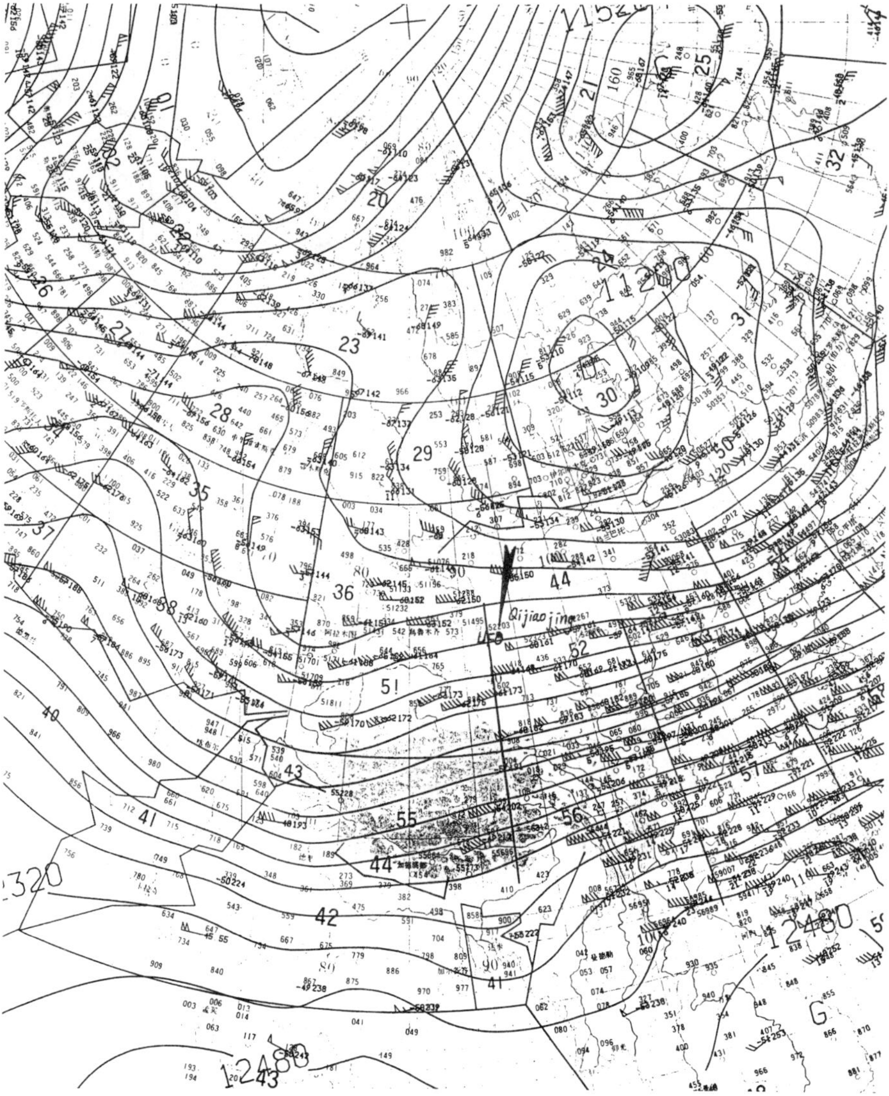

Fig. 2a. Air Flow Field at 200 hpa (≈ 12,500 m high) Over Xinjiang On March, 18, 1988 (data from the Central Meteorological Observatory of China).

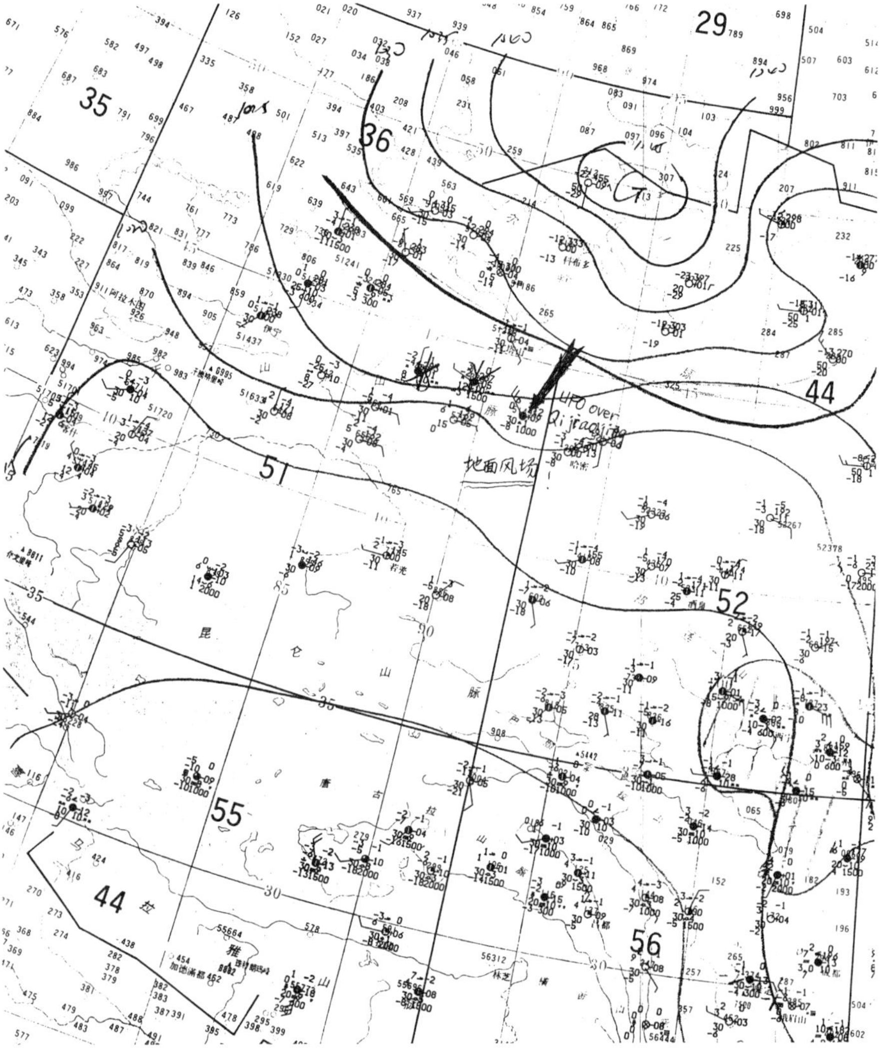

Fig. 2b Air Flow Field at Surface Ground (1000 hpa) Over Xinjiang On March 18, 1988.
(Comparing Fig. 2a with Fig. 2b, We can see the wind shear between 200 hpa and 1000 hpa when the UFO appeared. Data from the Central Meteorological Observatory of China.)

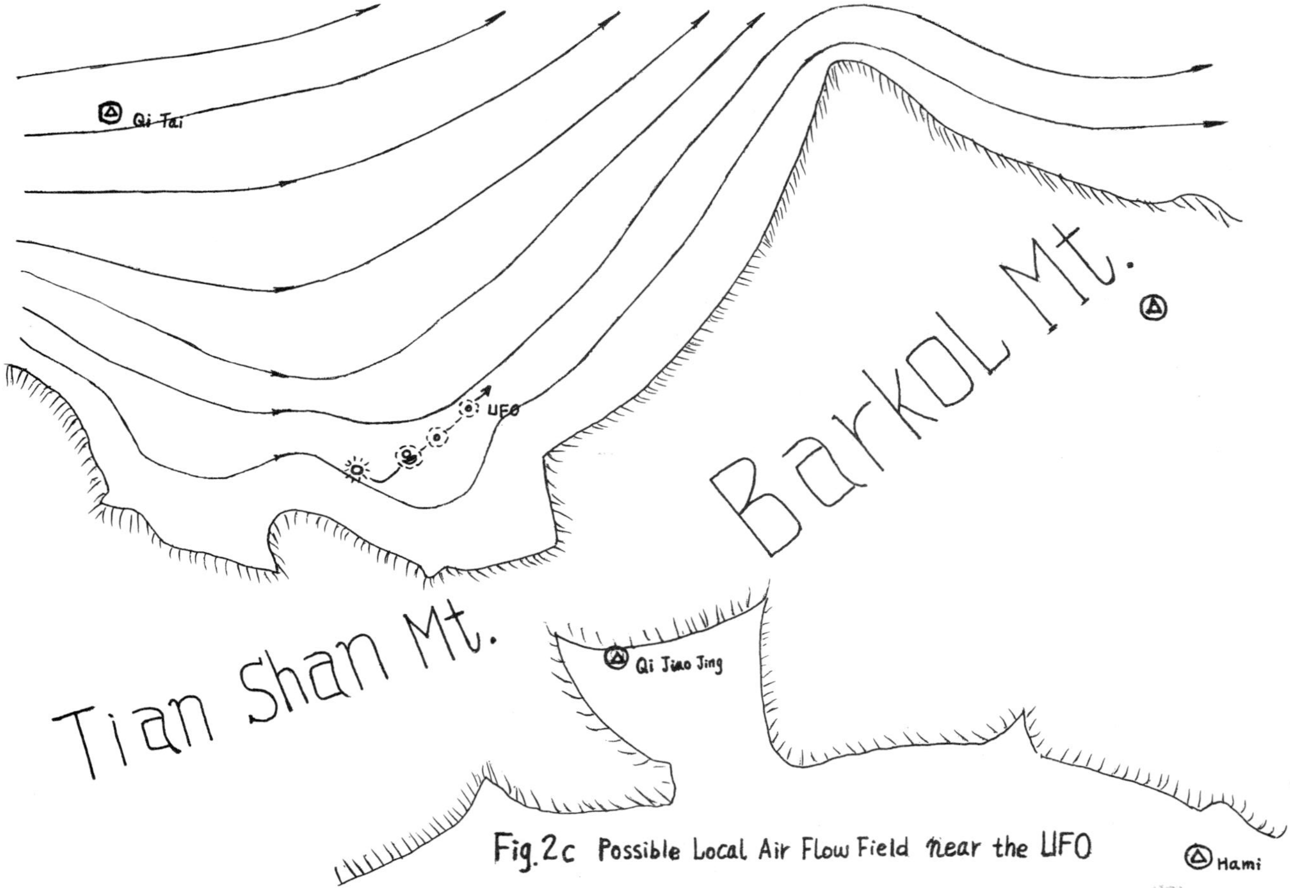

Fig. 2c Possible Local Air Flow Field near the UFO

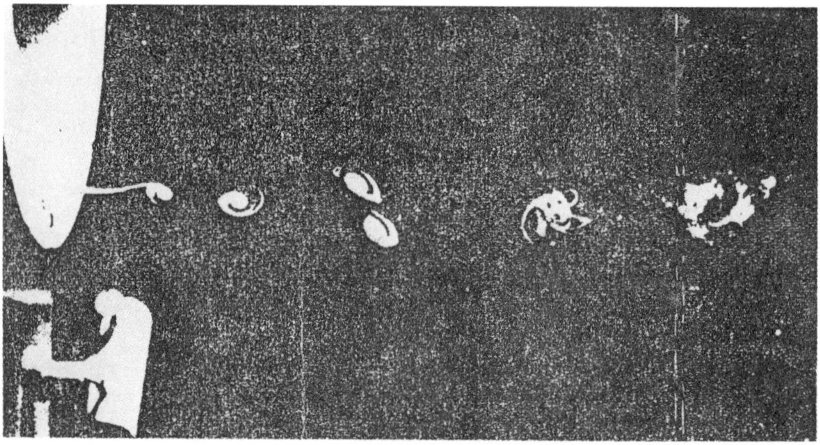

Fig.3. Instantanous photograph of rolling-up process of vortices, of slipping and coalescing of vortices, and of vortex disintegration. The smoke was introduced into the separated layer of an axisymmetric jet. D=7.5 cm(nozzle). U_0=3m/sec, f=95Hz.

P. L. Christiansen and P. S. Lomdahl/Numerical study of 2 + 1 dimensional sine-Gordon solitons

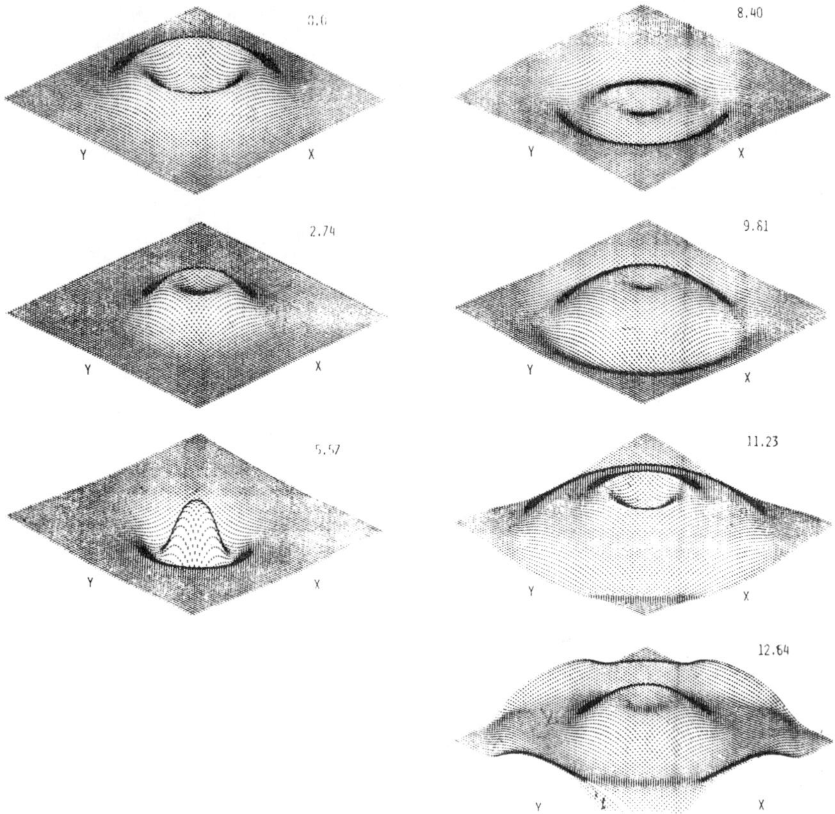

Fig.4. Soliton of sine-Gordon equation.

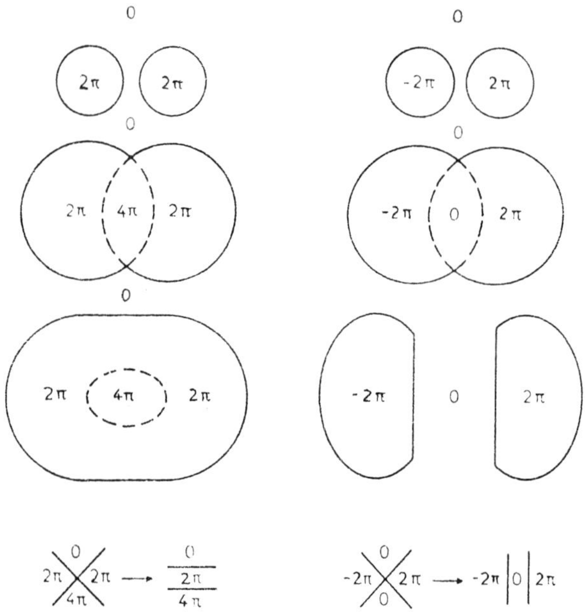

Fig.5. Sine-Gordon soliton`s slipping and coalescing.

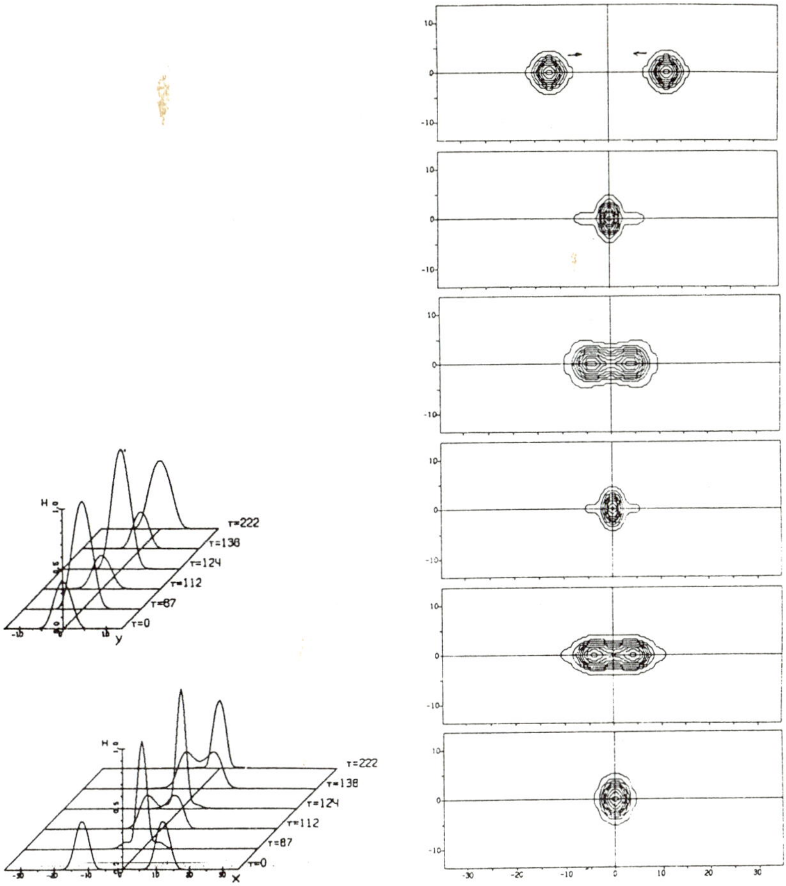

Fig.6. The interaction of two solitons in bound state (V.G. Makhankov, 1979).

Electric Reconnection, Critical Ionization Velocity, Ponderomotive Force, and Their Applications to Triggered and Ball Lightning

Hiroshi Kikuchi

Electrical discharge and lightning phenomena involve a large number of elementary processes competing with one another in a very complicated way. In order to highlight an overall physical picture, one should extract the most essential process and draw a simplified model. Surprisingly, however, there have been very few attempts to simplify a system of electrical discharge science from the point of a unified clear view, although a variety of different approaches were devised for individual phenomena, for example, by means of theoretical modeling and computer simulation. In this sense, it is no exaggeration to say that the theory of lightning and discharge is still in a preliminary stage and demands a new development. To effect this, it seems that the following directions are required: (1) more positive applications of electrostatics and electromagnetics; (2) creation of a new theory of collision-dominant plasmas involving ionizations; and (3) creation of new electrohydrodynamics (EHD), or more generally, electro-magneto-hydrodynamics (EMHD) combining plasma physics and meteorology. Along this line, the present paper introduces three novel concepts, (1) electric reconnection; (2) critical velocity; (3) ponderomotive force that might play an essential role in linking the above three disciplines to lightning and discharge phenomena. Electric reconnection analogous to magnetic reconnection occurs in a natural environment, typically when a rocket-trailing wire is impinging on an electric cusp or electrically neutral region between the ground and two clouds (or double charge groups) with opposite polarities, causing electric field merging at the tip of the rocket. Consequent triggered lightning discharge may be considered a manifestation of electric field-line reconnection on the analogy of magnetic reconnection phenomena such as solar flares. The critical velocity concept for collision-dominant gases analogous to Alfvén's critical velocity provides a unified view of various ionization processes involved in natural, triggered, ball lightning, and laboratory discharges. Estimated values of the streamer, leader, and dart-leader velocities are in fairly consistent with observations. A new model based on the first two concepts sufficiently explains a bipolar flash in triggered lightning observed from winter thunderstorms in Japan. Similarly these three concepts seem to be relevant to ball lightning phenomena and are being applied to fire ball formation.

The PLASMAK™ Configuration and Ball Lightning

presented at the International Symposium on Ball Lightning; July 1988 (Tokyo, JAPAN)

P. M. Koloc, Prometheus II, Ltd.
Bx 222, College Park MD 20740
USA Telephone (301) 445-1075

The PLASMAK™ plasmoid (**PMK**), an ideally MHD stable plasma configuration, will account for many of the commonly observed characteristics which crop up repeatedly in eyewitness descriptions of the natural phenomenon Ball Lightning. The **PMK** model will be presented and examined in terms of these observed characteristics after a few preliminary comments.

Lightning Formation

For the purpose of understanding the nature of Ball and Bead Lightning and their formation, a brief review with an appropriate extension of lightning theory is necessary.

Over most non-mountainous terrain lightning strokes originate with a negative leader which establishes a narrow conducting channel to continually lower charge (electrons) as it propagates nearly to the ground. For the initial stroke this occurs in a series of relatively short steps. Branches often form, but usually one path dominates. A cylinder of charge is set up in the air as electrons is sprayed radially outward from the tip of the leader, due to the convergence of the electric field onto the channel tip. These electrons are then yanked outward to a distance of two or three decameters before they slow and attach to oxygen molecules. The slowing is related to the reduction in the field gradient due to an increase in radial distance from the tip and conducting channel. Once established, the charge cylinder shields the central charge supply channel from further field emission losses. The negatively charged molecules are quasi-stable and because their mass is much larger and diffusion rate is much lower, they effectively clamp further outward radial migration during the remainder of the lightning stroke.

The cloud potential is lowered by this process so that the vertical gradient field at the ground region under the descending channel increases sharply. When the vertical field reaches the breakdown voltage of two or three megavolts per meter, a positive streamer (arc) rises from the ground and meets the descending conducting channel at a substantial fraction of lightspeed. This is depicted in Figure 1.

From this point the description varies from currently accepted hypotheses regarding the nature and dynamics of the lightning discharge.

In presently accepted theory, the stored electric field energy is converted to the characteristic light flash and thunder through ohmic heating of the channel and subsequent expansion of the channel air by the discharge current of the lightning stroke. Unfortunately, this

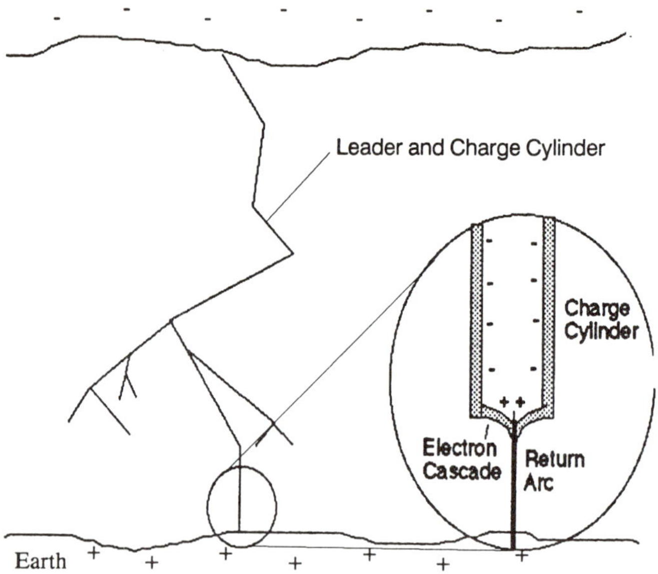

Figure 1 Common Lightning

does not account for the nearly total conversion (99%) of energy into the shock or acoustical components of lightning and the exceptionally low transformation of energy to the flash or light (less than one percent).

It is the author's contention that the principal energy conversion mechanism is current related but is NOT ohmic heating. Instead, it is a very efficient magnetohydrodynamic (MHD) process in which currents are temporarily carried by energetic (relativistic) electrons so only a very negligible amount of ohmic heating occurs.

In this scenario, at the instant the return stroke reaches the descending tip of a leader during one of its excursions and begins tracking back up the leader channel, the cloud to ground potential is reoriented locally to become a radial field with the field lines originating at the positive streamer and terminating on the charge cylinder. Because of the compressed distances and rapid closing speeds, the gradient radial electric field is an order of magnitude greater than what is necessary to establish an air breakdown (electron avalanche). Consequently, the charge cylinder electrons are ripped from their moorings on oxygen and other electrophilic molecules and are accelerated with greatly increasing speed between collisions. These electrons instantly run-away to relativistic velocities and are essentially collisionless for the rest of the trip inward to the positive streamer. They close and intercept the ascending return stroke. As the positive streamer progresses throughout the remaining charge cylinder and its branches, it collapses the charge cylinder and collects charge in a sequence like falling dominoes. As the electrons in each vertical increment section of the charge cylinder (ring) are discharged in turn, they follow a very low inductance disk like path to the anode streamer. Consequently, the rise time for lightning has been observed in the tens of nanoseconds regime, which corresponds nicely with the flight time.

The key to lightning channel formation of Ball and Bead Lightning has to do with the magnetohydrodynamics occurring for the few milliseconds beginning with the time of the closing of the return stroke and descending leader.

As each vertical increment of the more or less axi-symmetric charge cylinder closes upon the advancing return stroke, the energetic runaway electrons intercept and turn down the channel tip. This generates a flying tip of channel segment with a colossal increase in current and an explosive production of a dense surrounding azimuthal magnetic field. Since the adjacent air was preionized by collapsing electrons from the charge cylinder the instant before, and the rise time and field density are so extreme, an accelerating EMF sets up a transient one to one image current (coaxial transformer) immediately outside the surface of the stroke channel. This forms a conducting sheath that totally traps the developing azimuthal field. Since the inductance of the channel segment is large with respect to the inductance of the disk path, the driving EMF for the sheath current also causes its electrons to runaway to relativistic velocity. That is to say the reflected voltage driving currents in the sheath are of the same magnitude as the channel. The initial sheath current is substantially equal to the channel current.

Since each meter of channel length stores a megajoule of electric field energy or so, and the resistivity is essentially zero due to relativistic currents, a megajoule of magnetic energy is generated in the space within the sheath. At the outset of sheath formation the initial volume between the sheath and channel is zero. Within a few nanoseconds the currents could rise to the point where the field exceeds five kilogauss and the displacement of the sheath begins. The megajoule of magnetic energy exerts pressure which converts most of its energy to kinetic energy of the sheath. Once accelerated, the sheath ballistically moves radially outward while at the same time a slight compression displacement of the channel inward occurs which compresses its trapped axial field. This is shown in a cartoon in Figure 2.

With an initial impulse to fractional lightspeed, the sheath expands, picking up mass in a snow plow-like fashion, and decelerates to the point where the shock front no longer generates light. As the process continues the shock degrades to ordinary acoustic energy (thunder). Photographs of lightning bolts are probably picturing the light generating shock and not the channel. The shock produced light width could be a meter or so, while the return stroke channel is only a few centimeters at most. As the sheath expands and the vacuum field volume increases, a retarding EMF is generated which slows energetic electrons to more thermal velocities. Consequently, energetic electrons are not measured at the ground. Since the conducting sheath shields the channel currents at the outset, the peak currents in a lightning channel may not be correctly known.

The pinch effect in lightning has been considered a possibility for bead lightning, and it has even been claimed to have been observed. This is not likely. The earth's magnetic field

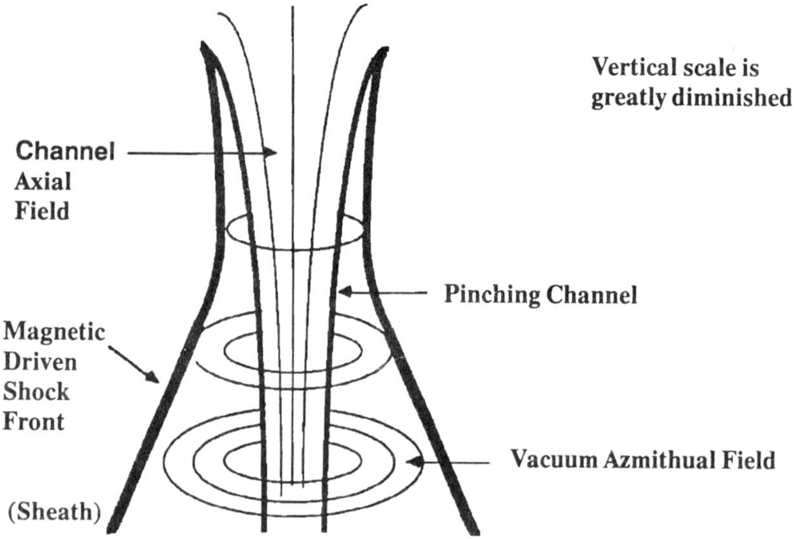

Figure 2 Magnetic Phase & Driven Shock Wave

is quite weak and has a sizable declination at a large range of latitudes away from the equator. The amount of this flux, which threads a circular cross-sectional area of a vertical charge cylinder (as formed by a cloud to ground stroke), is sizable. As the charge cylinder is discharged, the runaway currents flow across this cross-section to the centrally located channel. But, because of their high conductivity and fast rise time, these runaways compress and trap this threaded flux from the charge cylinder cross-section area into the cross-sectional area of the central stroke channel. See again Figure 2. Only vertical strokes near the equator or slant strokes which form a right angle with the earth's field will have large portions of their channels which do not envelop a substantial axial field.

The "pinch effect" in which a self generated azimuthal field squeezes and sometimes momentarily severs a plasma channel, cannot happen if that channel has a significant internal axial field. Pictures taken of possible lightning channel pinches may instead be narrow obscurations of the channel by tree limbs or power lines and the like. Incidentally, if the channel trapped axial field is quite sizable, the stroke will have some resistance to bending. For vertical strokes, the larger the earth's field declination, the stronger the axial field can be. Perhaps that may explain the [illegible]e looping forms of below the cloud main trunks of lightning that often are seen in the thun[illegible] belt of Florida as opposed to the more straight, milder zig zags and more direct paths to gro[illegible]d of lightning in Montana.

Bead and Ball Lightning

Now all the elements are present to see how lightning might form Bead or Ball lightning, or neither.

As an employee for Naval Ordnance Laboratory, the author had access to the motion picture of the accidental explosively triggered lightning event which terminated in the formation of a spectacular string of lightning beads. Other such events have happened (Fieux, et al., 1975). When first discernible, they ranged from a decimeter to no more than a meter and one half in size. In each of the following frames the beads shrunk rapidly in size, with the smallest disappearing within a quarter second and the largest disappearing in not longer than five eighths of a second. A number of them seemed to be associated with the turning points of the initial stroke channel. A series of every third frame is shown in Figure 3.

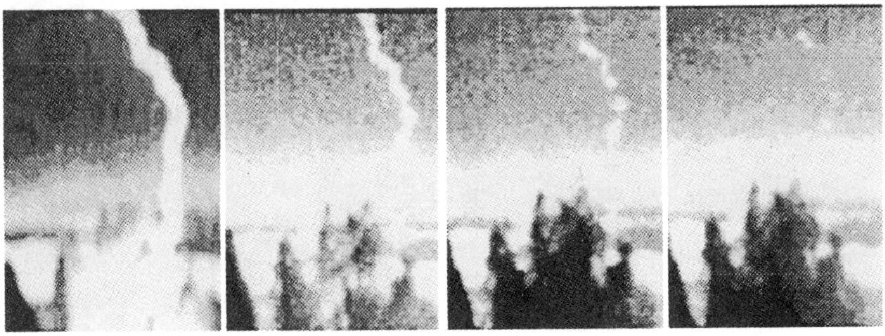

(a) (b) (c) (d)

Figure 3 A Bead Lightning Event

At the time that the bolt occurred, explosive shocks had already overlapped from three approximately equally distant (measured in the plane of the water's surface) break out points during an underwater mine test. The sudden raising of the salt water surface probably triggered the bolt. The overlapping of the three shocks created an intense acoustic air density interference pattern which formed a preferred series of twisted rarefactions through which the channel was seduced. Several of these were helical. Thus the channel contained several spots of "lumped inductance". Since the total series inductance was large, the channel rise time was greatly slowed. Consequently, more energy was stored in the lumped inductors and less in the driven shock. Under MHD effects the helical section coalesced to form closed rings or toroids. With the lack of energetic electrons, the resistive decay time for these magneto plasmoids was only a few tens of microseconds, but since the energy along the line was concentrated at these points, much greater heating was available to the locally surrounding air. This resulted in the solid plasma balls of various sizes that were captured by the motion picture camera in later frames. Bead lightning is very beautiful, and the beads observed on another occasion by the author were dazzling in their brilliance.

Ball Lightning

If there is a single (or very few) occurrence(s) of (a) helical section(s) along the lightning channel, then the overall channel inductance will still be very low and the rise time sufficiently fast to spawn energetic electrons and an excellently formed sheath, as first described above.

Due to the additional inductance per unit vertical length of channel, a segment of the lightning channel in a helical shape will generate more local flux just external to the channel. The added energy per unit length of channel is entirely due to the poloidal component of field resulting from the circular flow of current following the helical path. The plasma jacket or sheath surrounding the vacuum field of the helical channel segment is ballooned outward by the associated increase in field pressure. This forms an outward bulge in the otherwise cylindrically accelerating sheath. This is shown in Figure 4. The ballooned sheath and attached unballooned channel sheath may have been photographed as reported by Petrzilka (1967).

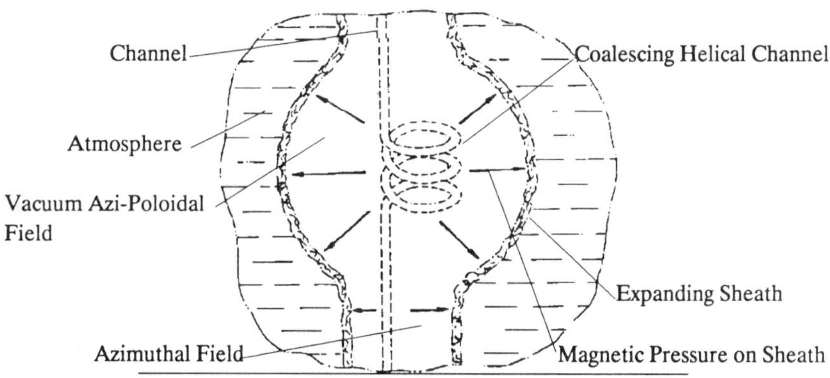

Figure 4 Forming Ball Lightning

As the sheath expands, the helical stroke channel also changes dynamically. It has an accelerating contraction along its major axis, and tends to expand in the axi-normal mid-plane. Within tens of microseconds the plasma loops contract to the point where they coalesce and form a toroidal loop. The currents and internally trapped axial field reconnect during coalescence to form a plasma ring which can relax to the Taylor state, containing both toroidal and poloidal components of current and field. The image current in the sheath responds to the change in helicity (pitch of the flux lines) and generates a counter helical (barber-pole) image current in response. When the coalescence is complete and the remnant axial lightning current in the channel is extinguished, the innermost edge of the sheath forms a current which is anti-rotational to the toroidal component of current in the plasma torus. There is no image current to the internal toroidal field of the ring since it does not and never did it, at an earlier morphology, impinge upon the highly conducting inner surface of the sheath.

Finally, with the azimuthal field dissipated, the cylindrical portions of the stroke collapse and extinguish, leaving a spheroidal magnetoplasmoid suspended in air continuing to exist off its considerable magnetic energy. The Mantle and insulating vacuum field protects the Kernel plasma, and in turn its high temperature cooling radiation maintains the necessary ionization in the inner portions of the Mantle. In the case of Bead Lightning the excessively slow rise time diminishes or precludes the formation of an energetic current sheath, so that this sheath would disintegrate, allowing atmospheric air to pour in and cool and release any otherwise trapped energetic currents. Without this conductivity the magnetic life time is reduced by five or six orders of magnitude. That converts magnetic energy into more and hotter plasma via catastrophic ohmic heating. In a well formed Ball Lightning, stability is aided by the alignment of currents and field lines in the Kernel plasma, the ultra high conductivity of the energetic currents, magnetic shear, and a uniform pressure over the external flux surface of the poloidal field. This latter condition makes the object omnigenous, and that condition reduces energy losses as well as aids stability. The model being proposed here is that of the PLASMAK™ configuration.

The PLASMAK™ configuration

The quest for fusion forced science to recognize the views of ancient Greeks that four states of matter exist, earth (solid), liquid (water), gaseous (air) and fire (plasma), instead of just the first three. Fusion research may have been a ploy used by some to seek easy funds from their governments to develop plasma science (Heppenheimer, 1984). Certainly, by far the major portion of funds spent in plasma physics has been tied to fusion research, and to a lesser degree to space plasma physics. Secret investigations of very dense plasma states associated with nuclear and thermonuclear weapons have also been funded, but such information tends to be esoteric. Funding of plasma science concerned with the plasma regimes of atmospheric electricity has yet to be seriously considered.

An evolutionary path was struck by fusion plasma science in the development of magnetic fusion devices, especially the mirrors, stellarators and the tokamaks. The open mirror evolved to a closed stellarator. This toroidal device evolved to a tokamak by conversion of the toroidal coil currents to a current within the plasma doughnut. It is at this stage of development that the world effort is currently positioned. Funding of fusion is so large that simply to obtain slight scientific improvements in the major programs has placed almost all other programs and new startups under severe pressure. The international effort seems to have now collapsed back onto the 1960's tokamak (Figure 5).

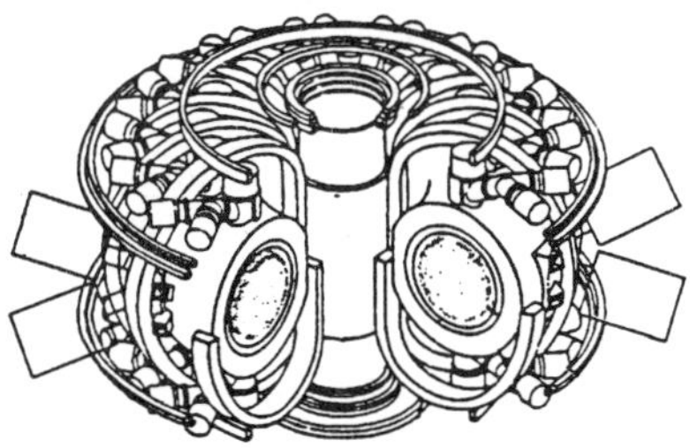

Figure 5 A Tokamak

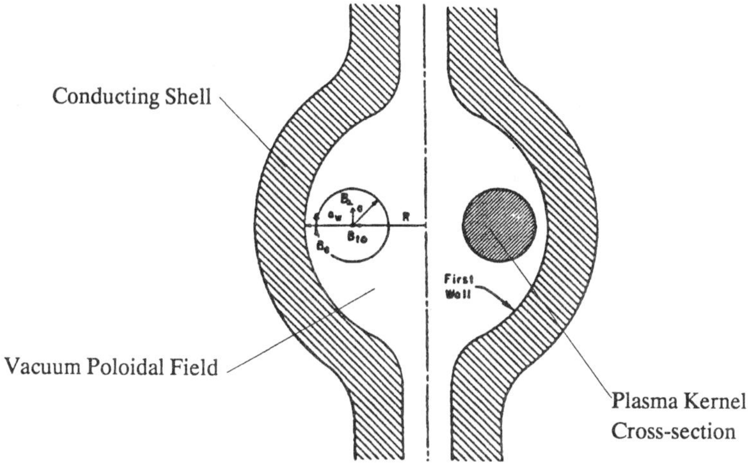

Figure 6 The Spheromak Configuration

The introduction of the Spheromak and PLASMAK™ configurations (Koloc, 1977; Rosenbluth and Bussac, 1979) has provided the next two evolutionary generations. A bit of work has been managed on the Spheromak in the United States and Japan (Katsurai, et al., 1980), and there is also some theoretical work in Europe and a sizable classified program in the USSR, a spinoff from the work of Kurtmullaev (1981). The Spheromak has both poloidal and toroidal currents within the ring; thus it eliminates the need for large toroidal field (tokamak-like) coils that link the plasma ring. The Spheromak is depicted in Figure 6; notice the similarity to Figure 4, which shows the plasma helix before coalescence and a plasma sheath instead of a rigid conducting shell. The principal advantage of a Spheromak is that the plasma pressure as measured on the toroidal axis is about ten or twelve times the confinement pressure at the conducting boundary. IF that boundary is generated with image currents in a highly conducting shell and NOT synthetically using magnetic coils, the Spheromak is ideally MHD stable. It was the fear of many for a long time that a stable magneto-plasma configuration could not be found.

The PLASMAK™ configuration is the omega point (final point) of this evolutionary track, since even the current bearing confining boundary is a plasma. That means that a PLASMAK™ magnetoplasmoid has reduced the complexity of a tokamak reactor device to a simple gas blanket supported magneto-plasma bubble. It has all the advantages of the Spheromak. The additional characteristic of this special plasmoid is that the plasma currents are carried by energetic relativistic electrons which give it exceptional conductivity, lifetime and improved stability.

Figure 7 depicts a copy of the Schneidermann photograph of a Ball Lightning as reported by Norinder (1965). A PLASMAK™ magneto-plasmoid is depicted in Figure 8; it

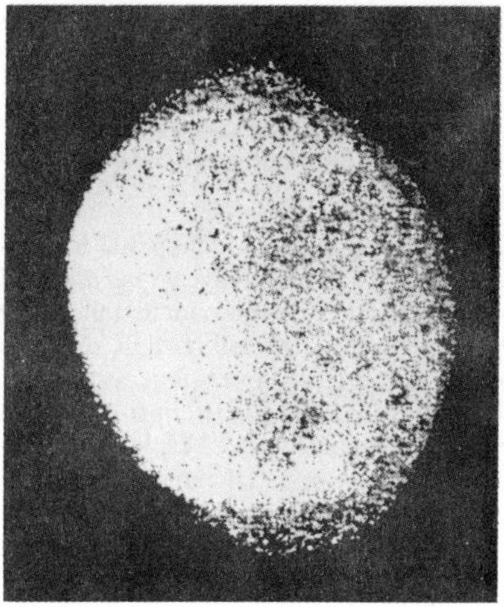

Figure 7 Natural Ball Lightning

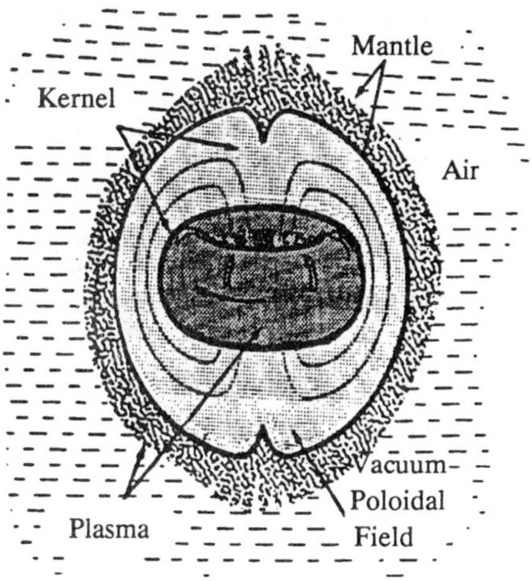

Figure 8 The PLASMAK™ Plasmoid

consists of a central plasma ring with its currents and fields as well as an insulating surrounding vacuum poloidal field that is trapped within a jacket or shell of plasma. The ring and its currents and field is referred to as the Kernel, and the highly conducting plasma shell with its energetic electrons skating on a vacuum field / plasma boundary is referred to as the Mantle (Koloc, 1980). Thus PLASMAK™ is a contraction of the phrase **PLAS**ma **M**antle **A**nd **K**ernel. To be brief, **PMK** refers to a PLASMAK™ magnetoplasmoid, and by extension to its naturally occurring version, Ball Lightning. The polar feature on the ball may or may not be observable depending on the amount of stray field (percentage of poloidal flux not trapped by the highly conducting Mantle) and the thickness of the Mantle, which can vary with the temperature (penetration power) of the bremsstrahlung (cooling) radiation from the Kernel plasma.

The commonly observed characteristics of Ball Lightning should be compatible with what is known about the **PMK**/Spheromak model. Some of these will now be examined, with reference made to excerpts from selected eyewitness accounts which are reported in examples or depicted in a few prints of photographs, sketches or paintings.

Energy

Since the PLASMAK™ configuration is a Spheromak topologically, the two have the same magnetic energy (Bussac, et al, 1978) and pressure distributions. The field just inside the Mantle must support the atmospheric pressure at equilibrium. That value is about five kilogauss, or half a Tesla. On the toroidal axis that value is sixteen or seventeen kilogauss.

The internal energy of a Ball Lightning could easily be four to ten times that of the equivalent atmosphere. This does not include energy stored in atomic states, ionization or chemistry.

> *During a thunderstorm I saw a large, red hot ball come down from the sky. It struck our house, cut the telephone wire, burnt the window frame, and then buried itself in a tub of water which was underneath. The water boiled for some minutes afterward, but when it was cool enough for me to search I could find nothing in it. (Morris, 1936)*

This report of a ball descending into a tub of rain water and boiling the water is interesting but not completely dependable since the size of the ball is not given. Further, the measurement of the temperature of the water by hand did not take into account that the water heated by the Ball Lightning encounter could have been a relatively small portion, since the heated water would have risen and remained very near the surface. A very thin evaporative cooled layer could have reduced bulk cooling at the surface leaving a very hot-to-the-touch layer just beneath. The "boiling" could have come from the highly localized explosive cavitation which would be similar to that generated by a red hot piece of metal plunged into water and forms a thin insulating vapor layer around its surface. Some of the gas released from the Ball Lightning under water could have come from gradual deionization of the Mantle plasma and its subsequent release as gas. Consequently, a more conservative estimate might consider only four liters to have been heated to boiling. This requires a little over a megajoule. If a Ball Lightning based on the PLASMAK™ model has two thirds of its energy in magnetic energy and the rest in atomic or chemical energy, then that Ball Lightning would have had a crudely estimated diameter of eighty centimeters.

Direct prolonged contact with the surface of a Ball Lightning can generate burns and heating. In Figure 12, a copy of the Bundy museum sketch of Ball Lightning clearly shows smoldering straw ignited by a very robust Ball Lightning.

Lifetime

In a number of well known studies the lifetime of Ball Lightning has been reported to be in the seconds to tens of seconds range. In the PLASMAK™ model, energetic currents of an MeV will produce seconds lifetimes, and ten MeV currents can generate tens of seconds lifetimes. This dependence of a **PMK**'s life time on the energy of its electron currents is proportional to the electron current's energy to the 3/2 power. Since the inductance of a **PMK** is proportional to size, it follows that, excluding premature destruction by disruptions, the lifetime of a well formed **PMK** is also at least proportional to its size. There may be an additional variance with size related to a yet to be determined correlation between the energy of the conducting electrons and size. It is not so obvious how for a constant resistivity, the resistance varies with size, since it depends upon the ratio of currents which are focused in at the plasma vacuum boundaries where the densities are very low to the currents flowing in the bulk plasma. This is because the thickness of these boundary layers (affecting current cross-section) varies with pressure or density-temperature and not size.

Certainly Ball Lightning persists longer than an air-plasma blob of a meter or so in diameter as seen from the observation of Bead Lightning (Figure 3). In Figure 9, a composite of the Schneidermann photograph is used to demonstrate a lower bound on lifetime, the

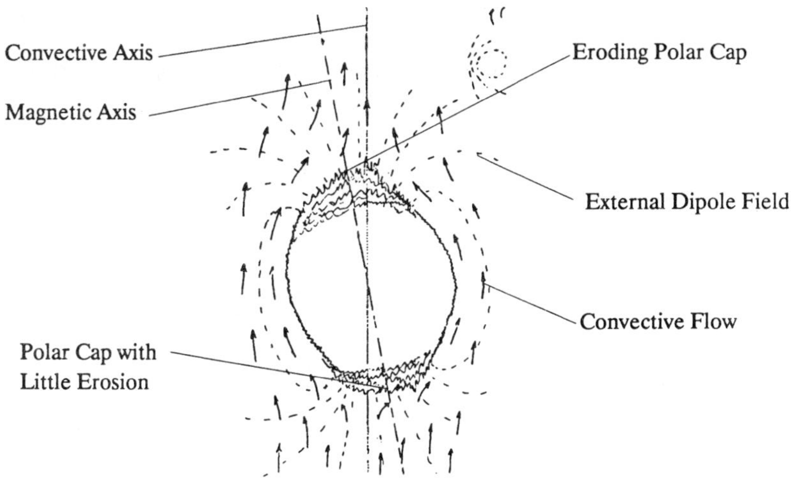

Figure 9 Evidence of Thermal and Magnetic Activity in Ball Lightning

elevated temperature of the surface, and magnetic characteristics of this Ball Lightning. The photo was taken in a North by Northwesterly direction in northern Europe where the earth's magnetic field has a declination of seventy to seventy five degrees. The Ball Lightning clearly shows the protuberances of the **PMK** model, and their axis is aligned with the earth's field. The sharp edge of the Mantle is caused by the clamping of electron thermal conduction by an embedded weak poloidal field parallel to the surface everywhere except at the polar caps. At the caps the thermal electrons are free to transport heat out along field lines with a large component normal to the surface of the Mantle. This is actually helpful since it allows the Mantle and field topology to act as a natural divertor.

In addition, it is clear that along the vertical axis a column of air is rising which flows over the surface and gains velocity with height so that it erodes the upper protuberance. This is due to thermal convection. The lower one is not eroded since the velocity of flow at the lowest extremity is essentially zero. This demonstrates thermal activity caused by a hot or very warm surface, and the time that it takes to establish such a flow, several seconds, shows that the Ball Lightning must have been in existence for at least that long.

Magnetic Activity

An historical observation of a woman who was "chased" by a Ball Lightning at her own wedding was cited in an ABC television program by Arthur C. Clarke and narrated by George Scott during 1984. The Ball Lightning "chase" in this recounting could have resulted from the near neutral buoyancy of the plasmoid and the natural tendency for light, airy objects to follow air currents flowing from room to room as they course in one side of a house and out the other. It would seem likely that the portals were open to ensure adequate ventilation for the guests. Additionally, considering the times and circumstance, it is likely that iron (magnetic materials) in the form of dressing pins or "stays" and wires in the supporting part of her wedding attire, including a fancy head dress, could have had an attractive effect on the weak stray field of the Ball Lightning, which is, by this model, a mild magnet. So if the theory is correct, Ball Lightning may have been displaying its natural "magnetic attraction" for the lady.

Occasionally Ball Lightning has been observed to roll, spin and "hop" over the ground. The presence of magnetic materials, such as embedded nails in flooring or ferrite particles in the soil, could account for its general attachment to the surface, and the variant dispersal of these materials can explain deviations in motion. In a **PMK** all of these effects would be due to field line tying with its external or stray field. **A PMK** would act as if it were magnetized, as it is.

Professor Richman was a scientist visiting Russia from the British Royal Academy for the purpose of collaborating in studies related to atmospheric electricity. Dr. Richman was using a long rod or pole "antenna" which penetrated through a hole cut in the roof so as to collect and to measure the changing electric field as low clouds would pass overhead. The galvanometer apparatus was used to measure currents in a gap in the pole about table level, while the "low" side of the pole on the other side of the gap passed directly into the ground. A lethal Ball Lightning was formed in the gap apparently when a lightning bolt struck the upper exposed end of the pole, and the gap apparatus was vaporized as the whole assembly momentarily assumed the function of lightning rod. Unfortunately for Professor Richman, the stray magnetic field of the Ball Lightning was probably quite large because of a slowly or poorly formed plasma Mantle (the external field trapping plasma conducting shell). It was observed to "bounce" off from his hand or wrist, and then the deflected ball struck his forehead and ex-

Figure 10 A Lethal Ball Lightning Event

ploded, thus rendering Professor Richman instantaneously and fatally unconscious. Figure 10 is a copy of a sketch of the event.

The physical concussion generated by the explosion was judged by a nearby assistant to have been the likely direct cause of Professor Richman's death. However, it is even more likely that the Ball Lightning's catastrophic disruption generated a lethal and internal brain stunning electric current due to a transient electromagnetic pulse. By the PLASMAK™ model this happened when the normally trapped and focused energetic currents of the Ball Lightning were spilled out through a rip in the conducting Mantle boundary. Because of the million times shorter "L/R" lifetime of the remaining magnetic energy driving replacement thermal currents, the field collapsed in a microsecond or two, all of which generated a lethal current pulse in the partially conducting brain. The original disruption or tear could have resulted from impact with Professor Richman's head or glasses. Proximity was no doubt a principal contributor to the dose of current received by the respective collaborators present that fateful day. More is discussed on the energy and explosive decay of Ball Lightning below.

Helicity of lightning driven fast rising currents in a galvanometer and its associated fields probably "induced" or plasmatized to form the culprit Ball Lightning. Ball lightnings have been formed by switch gear and may be of this same mechanism. This is discussed further below.

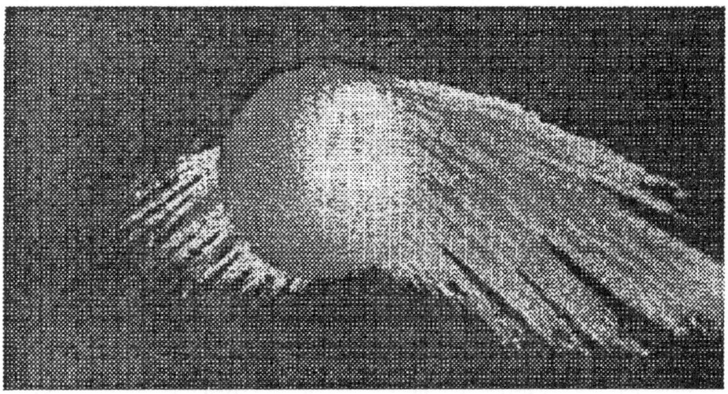

Figure 11 Rayed Ball Lightning

Rays and Energetic Currents

Figure 11 is a black and white copy of a painting rendered by an artist from his own memory which clearly shows a Ball Lightning at the moment of disruption. It shows two tangential sheets or beams pouring out in opposing directions from a longitudinal rent in the Mantle. The longer beam corresponds to dumping of the Mantle energetic electron sheet that is focused at the vacuum field-Mantle boundary. Here the electron currents confine most of the magnetic energy internally within the Ball Lightning and maintain a net longitudinal orbit as they do so. The focusing at the boundary comes from a balance between inward electric field pressure by a sharp boundary sheath that forms at the vacuum field-plasma interface, and the outward pressure of the magnetic field. Of course, overall inward force balance is achieved by the net inward force of the toroidal field with plasma ring acting upon the vacuum flux as it links through the "hole" of the ring. In addition the external pressure of the atmosphere results in fluid pressure through the plasma layer to the field via the current/fields interaction. The painting also depicts the much shorter but oppositely directed sheet of ions (probably mixed species) which are much heavier with opposite charge and circulation direction to electrons. Such an observation (as well as other "beamed Ball Lightning") can provide some important estimates on the energetics of the currents in Ball Lightning. This can be done assuming that the runaways, once released, will travel ballistically and slow as any dense injected beam of similar energy in atmospheric air. Inside the **PMK**, these electrons are COLLECTIVELY accelerated and maintained by the gradually decaying magnetic moment.

> *"November 24, 1886. Maracaibo, Venezuela.... The occupants... threw themselves on their knees and commenced to pray, but their devotions were almost immediately interrupted by violent vomitings, and extensive swellings commenced to appear in the upper parts of their bodies, this being particularly noticeable about the face and lips... The next morning the swellings had subsided, leaving upon the face and body large blotches. No special pain was felt until the ninth day, when the skin peeled off, and these blotches were transformed into virulent raw sores. The hair of the head fell off upon the side which happened to be underneath when the phenomenon occurred, the same side of the body being, in all nine cases, the more seriously injured. ...Another curious attendant circumstance is that the trees around the house showed no signs of injury until the ninth day, when they suddenly withered, almost simultaneous with the development of the sores upon the bodies of the occupants of the house." (Cowgill, 1886)*

E. T. F. Ashby related a story which made the rounds of the fusion community that he had found evidence of induced radioactivity from collision and disruption of a Ball Lightning with a fireplace in a homestead in the United Kingdom. Radioactivity can easily be induced by relativistic currents above five million electronvolts. Estimates for a larger air borne **PMK** is that it could trap relativistic currents with energies an order of magnitude higher than that threshold. In the account above evidence for soft X-rays is given, although the damage could have been caused by nuclear transmutation by the electron currents captured by heavy ions that found their way into the plasma Kernel, perhaps at formation.

Color, Hue, and Odor

Other external observations that give clues include such incidentals as color, and the change in hue and brightness can signal its changing dynamic state. Some have halos of purple or blues that are noticeable when the background illumination is subdued. Such halos are caused by the photoexcitation of extreme ultraviolet radiation that penetrates the air for one to three centimeters.

Lightning Balls that float quite freely in the atmosphere are commonly yellowish to red orange. This is because copious amounts of nitrogen oxides and ozone are generated by combined effects of ionizing radiation from the hot Kernel plasma ring and energetic currents at the inner Mantle edge, and these in turn can produce other molecular complexes including the explosive gas nitrogen pentoxide. These gases easily capture stray electrons, become negative molecules and can "stick" to the surface magnetic field and form a dense, blue and green light-absorbing layer. Combinations of ozone and nitrogen oxides can render a "sulfurous" odor which has been commonly reported.

Viewing optical radiation of a Ball Lightning coated in these gases filters the light, rendering a range of hues from yellowish to deep crimson. If the Ball Lightning is magnetically coupled to its local environ, this layer can be "blown away" by a relatively gentle air current. In very still air a Ball Lightning has been seen to have a line of reddish gas (probably nitrogen pentoxide) streaming downward or "dripping" from its lowest extremity. In some cases Ball Lightning plasmoids have brushed organic matter which has altered its color to a pinkish-red, probably due to the emissions from excited carbon. Green Ball Lightning has been generated by aberrant discharges across copper electrodes or wire bus involving high current or large inductive electrical equipment (so-called Switching Gear Plasmoids). The observation that Ball Lightning generates O_3 and NO_x has been observed scientifically, as noted in the account of a very fortunate sighting given below.

> *"An observation reported in unique detail was made in 1967 by Dmitriev, a chemist with experience in plasma studies. The ball appeared after an intense flash of ordinary lightning and passed over the observer, who was camped on the banks of the Onega River in Russia. . . The glowing mass contained a bright yellow-white center 6 to 8 cm in diameter surrounded by two outer shells. Next to the bright core was a dark violet layer 1 to 2 cm thick, and outside this lay a bright blue shell 2 cm thick forming the outer boundary of the ball. The entire mass was somewhat oval in shape, a few centimeters longer than it was wide. . . The witness reported taking four samples of air soon after the ball passed over him with evacuated gas sample bulbs which happened to be ready for use. Analysis of the gas collected indicated the presence of both ozone and nitrogen dioxide in amounts much greater than normal in air... Other gases usually found in air were unchanged in concentration. The observer noted a strong odor at the time resembling that produced by high-energy radiation in air. A transistor radio in the observer's camp gave increasing static in the early stages of the incident, which soon became a continuous rumble of increasing loudness. The ball itself was heard crackling loudly at close distance by the eyewitness. It left a trail of bluish, acrid smoke which slowly dissipated. As it approached the thicker forest, the globe collided with several trees and emitted violent sparks in six or seven such encounters. The collisions caused it to zigzag, and its color changed from the original white to bright red. After remaining in view for 60 to 65 sec, the ball disappeared. (M.T. Dmitriev, 1969, as reported in Singer, 1971).*

Pulsating and Modulated Light

Many pictures exist of what appears to be a pulsing Ball Lightning in flight. These are probably time exposures of street light pulsing at one hundred twenty hertz (driven by 50/60 hertz AC), and the "flight" was caused when the camera was inadvertently moved before the shutter was closed. It seems unlikely that, if Ball Lightning can be explained by the PMK model, it would pulsate in such a way that its intensity would vary essentially by one hundred percent, as seems to be indicated in many of these photographs as presented. Further, such modulation has not been reported by direct human observation. That does not mean that a regular modulation (probably much slower) but at a fraction of full light intensity does not occur. This can happen if the Kernel plasma ring is precessing or tumbling, as could occur with line tying of stray magnetic flux to a background earth's field, or other external and especially varying magnetic influences. This particular effect may be more noticeable if viewing the plasmoid through polarizing filters, as commonly found in certain types of sun glasses, or from partially reflecting surfaces. This is expected because some of the source radiation will be polarized by field effects.

External Power Supply?

The above cited ABC TV program also included an interview with an observer of Ball Lightning on board an Eastern Airlines flight, the British radio astronomer R. C. Jennison, and it is typical of many such observations and demonstrates a number of unique properties. The Ball Lightning's external stray field would naturally set up image currents in the conducting aluminum skin of the aircraft as it drifted, driven by front to aft air circulation from one end of the aircraft to the other. The resultant slight pressure caused by compression of the conserved stray field flux by the aircrafts metal skin would be all that is needed to orient the ball on the axis of the aircraft. Since the Ball Lightning was stabilized against lateral displacements of air flow, and mild air flow over the Ball Lightning's surface would result, this would wipe off the normally present buildup of the light screening nitrous oxide blanket. Consequently, this Ball Lightning, instead of having a yellowish or orangish hue, in this case was a beautiful bluish color.

Professor Jennison referred to the "SCREENED" [meaning confinement within a highly conducting enclosure constructed of materials such as metal] aircraft, which was a scientifically significant statement. This is because a "screened" object can NOT depend on an external source feed for auxiliary electrical energy with which to maintain itself. Thus models of Ball Lightning requiring either or both external electromagnetic and electric fields for continued existence are eliminated from consideration. On the other hand, a PLASMAK™ plasmoid does not require continuous external energizing since, due to its high internal energy from the outset, it has sufficient energy to meet all but the most stringent reported Ball Lightning feats.

Brisance

The ABC TV program cited above also reported a Florida observation of Ball Lightning that was a somewhat larger than average orange ball with bluish and violet halos. The halo is probably due to air excitation by extreme ultraviolet generated by the recombination of N^{++} and O^{++} and an air penetration depth of one or two centimeters. The ball was reported to have been displaced to the floor when the observer swung her fly swatter at it. The force of the fly swatter probably created a localized down draft, or the iron wire in the mesh and handle attracted the Ball, which resulted in the Ball Lightning being pulled downward, where it momentarily "rolled around" and exploded with the intensity of a "shotgun" blast (very loud fire cracker).

It has been commonly observed that Ball Lightning terminates explosively perhaps fifty percent of the time, and in some cases the explosion has caused considerable physical damage. The scientific explanation of this is that Ball Lightning has a high internal energy and is disrupted by contact with the external environs or because the re-ionization power weakens to the point where neutrals invade and cause resistive disruptions. The **PMK** model contains about an order of magnitude higher average energy density than the surrounding air. Interdiction and loss of the energetic electron sheet current at the Mantle and vacuum field interface instantaneously drops the conductivity and causes thermalization of the stored magnetic energy within a few microseconds. This immediate conversion of the remaining magnetic energy shock heats the plasma, which is now no longer confined by the resistively consumed magnetic field. A supersonic expansion wave results, which develops a sharp front as it propagates through the portions of the Mantle containing nitrogen pentoxide or other unstable and labile photo-compounds, and triggers their explosive decomposition. This process enhances the brisance of the outgoing shock wave. Since this particular Ball Lightning (Florida incident) was orange, it probably did contain some peroxides of nitrogen.

Bounce

The generation of Ball Lightning on board electric submarines was observed when emergency bank switching was used to practice disaster diesel engine startup procedures. Since in this exercise the current switched exceeded the rated switch loading when all of the electric submarine batteries were switched simultaneously, the practice was only done once or twice each year. In one design the switch was in a compartment on one side of the conning tower and was operated by mechanical linkages from the compartment on the other side. A small tunnel or crawl space under the conning tower connected the two compartments. During the exercise, green Ball Lightning would form in the switch box and then travel through the crawl space and into the control compartment. The Ball Lightning would then sting the legs of the switch operators as the plasmoid extinguished with a pop, which is probably due to a milder form of the EMF or EMP effects that caused the death of Professor Richman. The greenish color was probably due to the excitation of neutral silver or copper vapor originating from electrode erosion during switching. It would seem that chemistry of the elements is not the important factor in Ball Lightning, but rather magnetoplasma effects.

James Tuck succeeded in producing and filming a Ball Lightning in the laboratory. The problem that Tuck faced in exploiting his work on Ball Lightning was that he was in his retirement phase-out period at Los Alamos when he started his project. At the time Ball Lightning was considered to be "impossible" by too optimistic an application of the "virial theorem" (Finkelstein, 1969), and therefore there was no follow up interest in restarting the work. The two rolls of film photographing the event were not developed until several months after Tuck had left the laboratory's employ in 1970 or 1971.

All but the last of his experiments did not include the arc suppression gear, which consisted of a large iron "C" shaped flux core ring with the switch placed in the gap of the ring. Such an apparatus has the magnetic helicity and can generate the EMF's necessary to form a **PMK** like Ball Lightning with energetic currents. Tuck's last experiment produced a ten centimeter diameter, very viable, and rugged Ball Lightning (J. Tuck, 1970), although this formation technique is obviously not efficient. The original film more clearly shows the ball bouncing when it intercepted the floor. Bouncing means that the internal pressures are larger than the surrounding atmosphere. It indicates, too, that the ball is super stable and tremendously rugged, as is predicted by the **PMK** model. Also, there was no evidence of a trail of metal oxide "smoke", as would be generated by a rapidly burning chunk of metal, i.e. magnesium. The ballistic path of the object did not show a "parabolic track" as would be the case for objects whose mass density was greatly larger than air. The rapidly moving object appeared to be bluish white, and as was expected did not appear to build up a surrounding nitrogen oxide and ozone gas blanket.

The Ball Lightning did not fade; it was vectored downward at an angle and bounced from the concrete floor like a superball, but its trajectory was quite straight with a curvature in the plane of the floor that seemed to be pulling it toward the steel wall to which it was traveling adjacent. Compression of the gas plasma boundary upon impact will cause deformation of the fields. Because the fields are more dense the nearer the Kernel plasma ring, the system is Rayleigh-Taylor stable. The flux, which is compressed more at one side of the Kernel, exerts more pressure so that the whole magnetoplasmoid bubble is repelled intact. Soap bubbles are also stable, but only slightly so by comparison to a PLASMAK™ magnetoplasmoid. Although the pressure is higher within a soap bubble due to the combined effect of the external atmospheric pressure and the surface tension, it is just barely so. The average energy of a **PMK** is many times the boundary fluid pressure.

The Future

Applications of PLASMAK™ technology to fusion are exciting. Simple gas compression to liquid densities of the blanket surrounding a fusion fueled **PMK** will assure a burn. Since temperatures required for the burning of aneutronic fuels can be reached, the fusion energy can be absorbed in the dense blanket, turning it into a plasma suitable for the generation of electric power using inductive MHD technology. Protium Boride isotope eleven can be made to burn in just a few milliseconds, so the electric generator would operate with continuous power at fifty or sixty hertz. Burn densities of tens of thousands of millions of kilowatts per cubic meter are possible. Of course, each **PMK** would be compressed to only one hundred cubic centimeters or so. From understanding the simple phenomenon of Ball Lightning, we may be on the threshold of a new beginning in which Mankind and his dominion is no longer chained by gravity but will be free to move among the stars.

In conclusion, the PLASMAK™ model of Ball Lightning would appear to be a promising candidate to explain Ball Lightning. However, the mystery of Ball Lightning will be truly solved when sufficient funding is available for experimentation based on various models including the one introduced here. Substantial funding must continue until a model is found that can produce artificial Ball Lightning repeatably and the results can be shown to be identical with the vast majority of reported observations.

My appreciation to R.M. Fujii and D. M. Cooper in getting this paper out in order to make an earlier flight schedule.

*PLASMAK™ is a trademark of Prometheus II, Ltd.

Globular Lightning.

The Burndy Library

Figure 12

References

M. Bussac, H. Furth, et al., "Low-Aspect Ratio Limit of the Toroidal Reactor: The Spheromak," IAEA CN-37, Innsbruck, 1978.

W. Corliss (ed.), Lightning, Auroras, Nocturnal Lights, and Related Luminous Phenomena, Glen Arm, MD: The Sourcebook Project, 1982.

W. Cowgill. "Curious Phenomenon in Venezuela", Scientific American, 55, 389, 1886.

M. Dmitriev, "Priroda Sharovoi Molnii" [Nature of Ball Lightning], Priroda 56, 6, 98, 1967.

R. Fieux, C. Gary, and P. Hubert, "Artificially Triggered Lightning above Land", Nature, 257, 212, 1975.

D. Finkelstein and J. Rubenstein, "Ball Lightning", Physical Review 135, 390, 1964.

T.A. Heppenheimer, The Man-Made Sun: The Quest for Fusion Power, Boston: Little, Brown and Company, 1984.

M. Katsurai and M. Yamada, "Conceptual Design Study of Spheromak Reactors", PPPL-1614, 1980.

P. Koloc and J. Ogden, "The All Plasma Spheromak: The PLASMAK," Proc. U .S.-Japan Joint Sym. on Compact Toruses and Energetic Particle Injection, Princeton Univ., 216, Dec. 1979 (No Relativistic Corrections).

P. Koloc and J. Ogden, "The PLASMAK: Its Unique Structure, the Mantle," Proc. Third Symposium on the Physics and Technology of Compact Toroids, LASL, LA-8700-C, 204, Dec. 1980.

P. Koloc, "Plasmak Foramation," Proc. of the U.S.-Japan Jt. Symposium on Compact Toroid Research, PPPL, 20-23 February 1984.

P. Koloc, U.S. Patent No. 4,023,065, "Method and Apparatus for Generating and Utilizing a Compound Plasma Configuration," issued 10 May 1977.

R. Kurtmullaev, et al., "Experiments in the 'Tor-Liner' Device," Proc. Tenth European Conf. on Controlled Fusion and Plasma Physics, Moscow, Vol. 1, L-5, 1981.

J.R. McNally, Jr., "Preliminary Report on Ball Lightning", Oak Ridge National Laboratory, ORNL-3938, UC-34-Phys., 1966.

G.T. Meaden, "Remarkable Occurrence of Unusually-Shaped Ball Lightning," 15 August 1975, Journel of Meteorology (U.K.), 1:310, 1976.

W. Morris, "A Thunderstorm Mystery", letters to the editor of the Daily Mail of London, Nov. 5, 1936.

H. Norinder, "Comments on Ball Lightning," Proceedings of the Third Intl. Conference on Atmosphere and Space Electricity, Montreux, 1963. Amsterdam: Elsevier, 1965.

V.A. Petrzilka. Ball Lightning. Prague: Inst. of Plasma Physics, Czechoslovak Academy of Science, 1967.

M. Rosenbluth and M. Bussac, " MHD Stability of Spheromak," Nuclear Fusion 19, 489, 1979.

V. Shafranov, "On Magnetohydrodynamical Equilibrium Configurations," J. Exptl. Theoret. Phys., 33, 710, 1957. Transl. in Soviet Phys.--JETP, 6, 545, 1958.

S. Singer, The Nature of Ball Lightning, New York: Plenum Press, 1971.

J.B. Taylor, "Relaxation of Toroidal Plasma and Generation of Reverse Magnetic Fields," Phys Rev Letters 33, 1139, 1974.

J. Tuck, "Ball Lightning, A Status Summary to November 1971," Los Alamos Scientific Laboratory, LA 4847-MS, US-34, December, 1971.

M. Uman, Lightning, McGraw-Hill, New York, 1969.

M. Uman, All About Lightning, New York: Dover Publications, Inc., 1986.

EXPERIMENTAL RESEARCH ON BALL LIGHTNING

H.Ofuruton

Tokyo Metropolitan College of Aeronautical Engineering,

53-1, Minamisenju 8, Arakawa-ku, Tokyo 116, Japan

and

Y.H.Ohtsuki

Department of Physics, Waseda University,

4-1, Okubo 3, Shinjuku-ku, Tokyo 169, Japan

Abstract Experiments on producing ball lightning were made with discharge in flammable gas and/or aerosol. A long lifetime (2 s) ball lightning was observed in 2.7% ethane and $100 cm^3$ cotton fibers.

INTRODUCTION

There are many observational reports of ball lightning.[1] However, most of the reports are not complete and there are not so many photographs of ball lightning, because ball lightning is a very unusual phenomenon and its life time is very short. For these reasons, the properties of ball lightning are not understood enough.

Many theoretical models of ball lightning have been published, for example, interference of electromagnetic waves (Kapitza),[2] cosmic rays of antimatters (Ashby and

Whitehead),[3] chemical reactions and aerogels (Smirnov),[4,5] aerosol plasma (Mukharev),[6] boson (Dijkhuis),[7] and maser (Handel).[8]

Many electric discharge experiments of ball lightning have been made. There are two types of the discharge experiments. One is the type of using natural lightning and the other is using laboratory artificial discharge. In the former type experiments, the most well-known was made by Richmann (1753).[1] He erected a long metal rod which penetrated through a hole laboratory roof into his and used an electrometer connected to it. His cooperator reported that a blue-white spherical ball passed from the metal rod connected to the electrometer in his hand to his forehead. He was killed by an electric shock of this experiment.

Many experiments with laboratory electric discharge have been made. For example, Barry made a discharge experiment in low density flammable gas at standard atmospheric pressure.[9,10] His apparatus consisted of a glass chamber(252 liters) and copper electrode. The high voltage discharge was obtained by 5μF capacitor and 10kV power supply. The experiments were started by introducing propane, mixing the atmosphere, and triggering the discharge. In the range of 5.0 vol%- 2.8 vol% of propane, normal combustion phenomena were observed. In the range of 1.8 vol% - 1.4 vol% of

propane, a small yellow-green ball of fire was formed. Its lifetime was 1 - 2 s. We cannot see the change of this ball, because only one monochromatic photograph was published in Barry's book.

In this paper, we reexamined his experiment and showed the change of the fire ball. However, we did not use propane for the safety, and used methane or ethane. Moreover we made experiments of discharge in the air containing cotton fibers as aerosol.

EXPERIMENT

Figure 1 shows the shape of the electrodes used. The electrodes were made of copper rods (3 mm in diameter) and the spark gap was 3 mm. The glass chamber with interior dimensions of 734×372×429 mm was used.

Figure 2 shows the electric circuit. (Capacitor A:0.2μF, Capacitor B:2.5μF, Capacitor C:4.5μF)

The first discharge experiment was made in the methane-air mixed gas (Capacitor A:0.2μF). The s was in the ethane-air mixed gas (Capacitor B+C:7μF). The third was in the air containing cotton fibers (Capacitor A:0.2μF). The last was in the air containing ethane and cotton fibers (Capacitor B+C:7μF).

All discharges of these experiments were made in the

atmospheric pressure. Before these four groups of experiments, discharge experiments in the normal air were done, and ordinary sparks were observed.

The changes of the state in the chamber were recorded with video cameras.

RESULTS and DISCUSSION

1) Methane-air mixture and ethane-air mixture

At 1% methane experiment, the spark was the same as in normal air. At 2% methane experiment, a spherical light (3 - 6 cm in diameter) was observed. At first the green light was observed which was caused by copper ions from the electrodes. After 0.26 s this light changed into an ordinary ignition, because the concentration of methane of the upper part of the chamber was high enough. Figure 3 shows the change of these luminous states.

At the 1.4% ethane experiments, an normal air spark was observed. At the 2.1% ethane experiment, an fire ball (about 4 cm in diameter) rose up from near the electrodes. The color of the fire ball was red, but at the first it contained green color. The life time of the fire ball was about 0.3 s. Figure 4 shows the change of these luminous states.

At each experiment of methane and ethane , a fire ball

was observed. However, the lifetime of the fire balls was very short.

2) Aerosol (cotton fibers) and air mixture

Cotton fibers (about 100 cm^3 in volume) were torn into small lumps and were put in the chamber. The lumps of the fibers were scattered in the chamber by compressed air.

After the discharge, about 3 cm in diameter white luminous ball was first produced, became smaller, and disappeared within 0.8 s after the discharge.

Figure 5 shows the change of the luminous ball. In this experiment with aerosol, the color and the shape of the ball were different from the cases of the gas experiments, and the lifetime became longer.

There is some probability of ignition, because cotton fibers is flammable. However, the color of the luminous ball was similar to the color of the normal air discharge, not similar to the color of ignition of cotton. When cotton fibers were put between two electrodes and discharged, these cotton fibers disappeared instantly. This luminous phenomenon should be treated as aerosol plasma theory.

3) Ethane, aerosol(cotton fibers), and air mixture

At the last experiment, 2.7% ethane and 100 cm^3 cotton fibers (torn into small lumps) were used. After the discharge two white luminous phenomena were produced near the electrodes. One phenomenon (about 5 cm in diameter) disappeared after 2 s of the spark. Another one (about 1 cm in diameter) moved downwards and upwards, and disappeared. The lifetime of the latter lightning matter was about 1 s.

These two luminous balls were recorded on two video tapes. Two video cameras were set at right angles. Figure 6-1 and 6-2 show the changes of these balls. Scanning lines of these video picture were little distorted by the electromagnetic effect of the spark or some other causes.

The lifetime was prolonged because of plasma and combustion. However, it is necessary to analyze which is a main cause of the long lifetime.

The experiments were done under the condition of 18 C in room temperature and 100% relative humidity.

CONCLUSION

As the results of Barry's propane experiment and our methane and ethane experiments, the best concentration for producing ball lightning was just under the lower limit of combustion.

The lifetime of the ball was 0.3 s in 2% methane. In the experiment using cotton fiber aerosol, the lifetime was 0.8 s, and using both ethane and cotton fiber aerosol, the lifetime was about 2 s.

In these experiment, humidity, distribution of temperature, and constitution of the electrodes are found to be important factors for the production of ball lightning. However, the reproducibility of these experiments is not good and consequently much more experiments should be done in order to analyze the effects of these factors.

ACKNOWLEDGMENT

We are deeply indebted to Prof. T.Ito and Prof. H.Kobayashi (Waseda Univ.) for their help in this work. We wish to thank Mr. H.Ara for preparing the apparatus.

REFERENCES

1 J.D.Barry : 'Ball Lightning and Bead Lightning' (1980) Plenum

2 P.L.Kapitza : Dokl.Akad.Nauk.101,245(1955) (in Russian) (English translation : 'Collected Papers of Kapitza' Vol.2, Ed.by D.Ter Haar, (1965) Pergamon)

3 D.E.T.F.Ashby and C.Whitehead : Nature 230,180(1970)

4 B.M.Smirnov : Sov.Phys.Tech.Phys.22,488(1977)

5 B.M.Smirnov : Sov.Phys.Usp.30,420(1987)

6 L.A.Mukharev : Sov.J.Commun.Tech.Electron.30,77(1985)

7 G.C.Dijkhuis : Nature 284,150(1980)

8 P.H. Handel : Proc.8th Internat.Conf.Atoms.Electr.

9 J.D.Barry : J.Atoms.Terr.Phys.29,1095(1967)

10 J.D.Barry : J.Atoms.Terr.Phys.30,313(1968)

Figure captions

Fig.1 Electrodes (3 mm in diameter, copper rods)

Fig.2 Electric circuit

Fig.3 Methane(2%)-air mixture (discharge voltage 8 kV)
(a) 0.066 s after discharge
(b) 0.099
(c) 0.132
(d) 0.264

Fig.4 Ethane(2.1%)-air mixture (discharge voltage 10 kV)
(a) 0.033 s after discharge
(b) 0.066
(c) 0.132

Fig.5 Cotton fiber-air mixture (discharge voltage 9 kV)
(a) 0.05 s after discharge
(b) 0.10
(c) 0.20

Fig.6-1 (left) and Fig.6-2 (right)
Ethane(2.7%), cotton fiber, and air mixture (discharge voltage 8 kV)
(a) 0.264 s after discharge
(b) 0.660
(c) 1.254

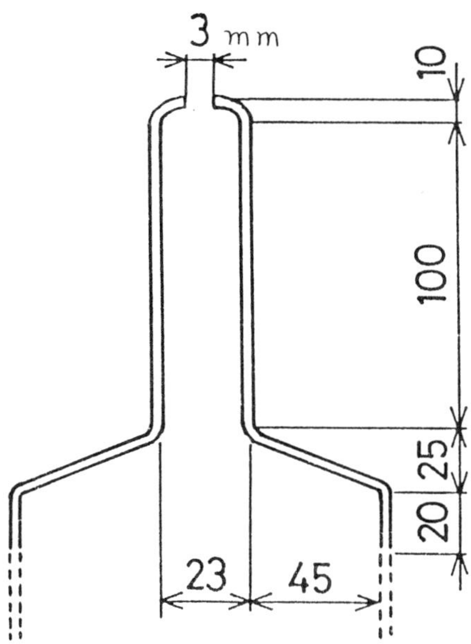

Fig. 1

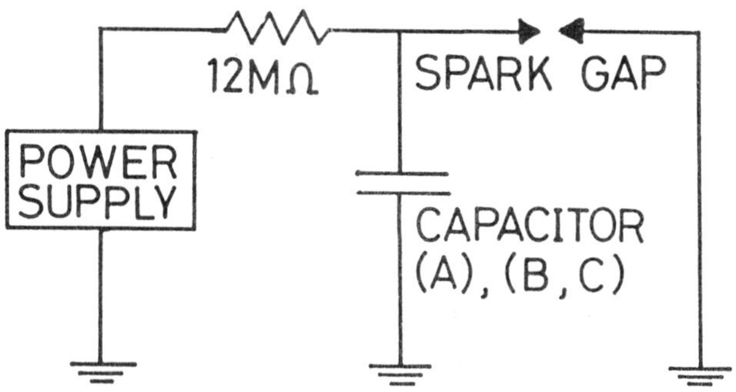

Fig.2

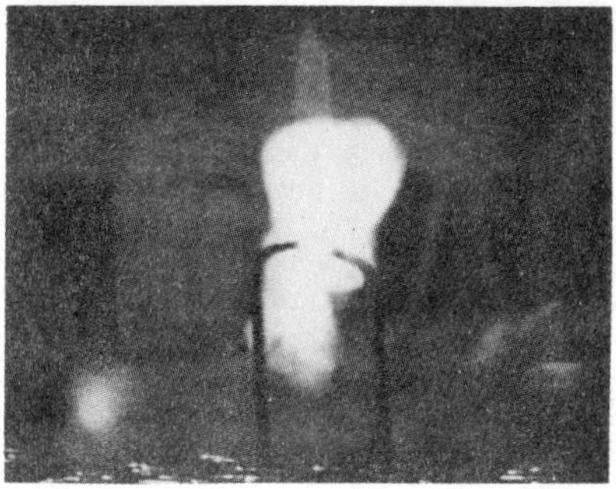

Fig.3 (a)

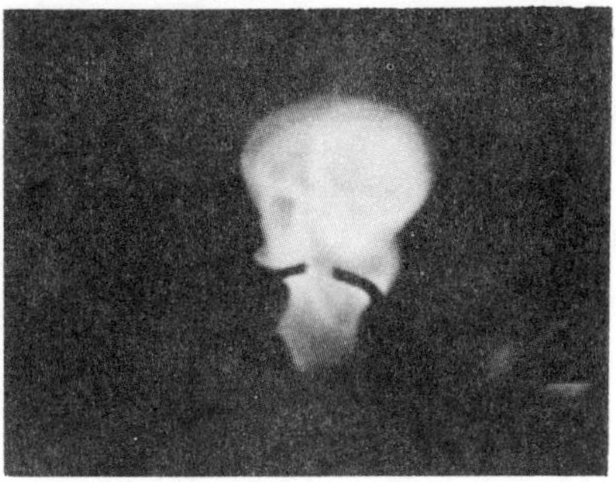

Fig.3 (b)

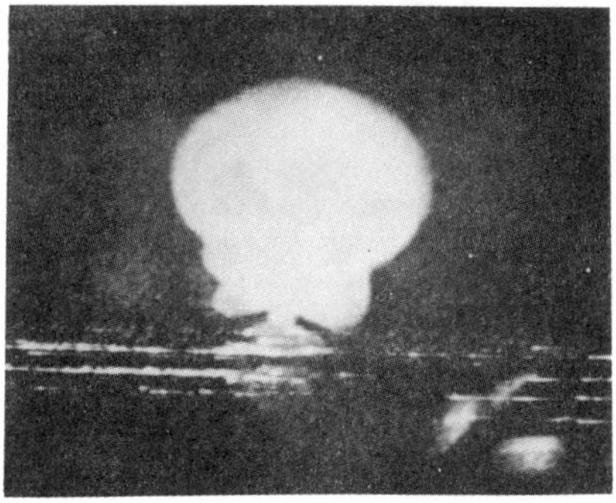

Fig.3 (c)

Fig.3 (d)

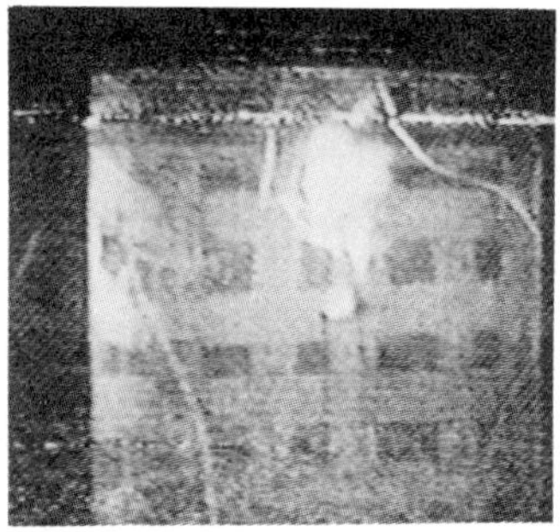

Fig.4 (a)

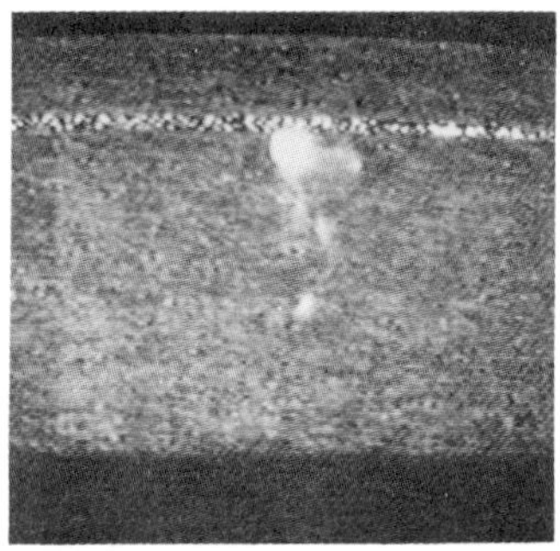

Fig.4 (b)

Fig.4 (c)

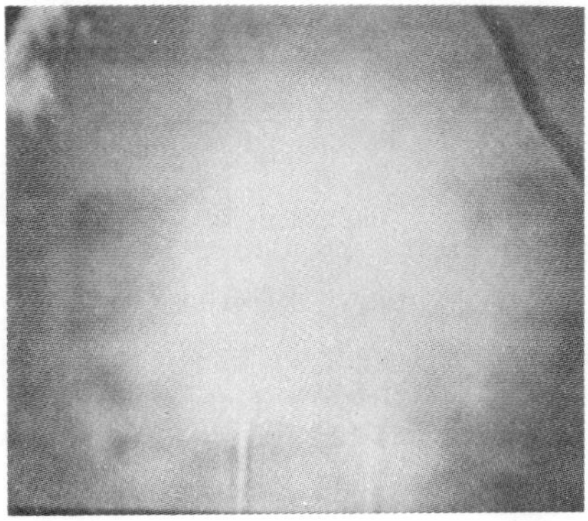

Fig.5 (a)

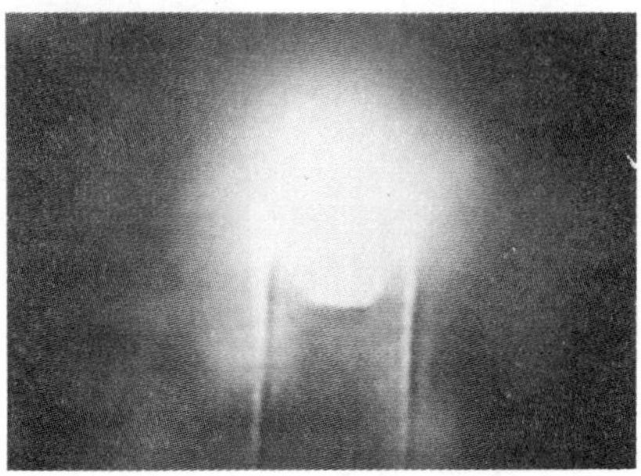

Fig.5 (b)

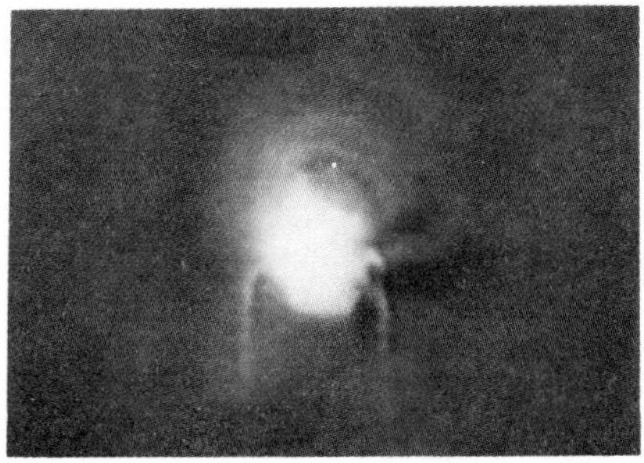

Fig.5 (c)

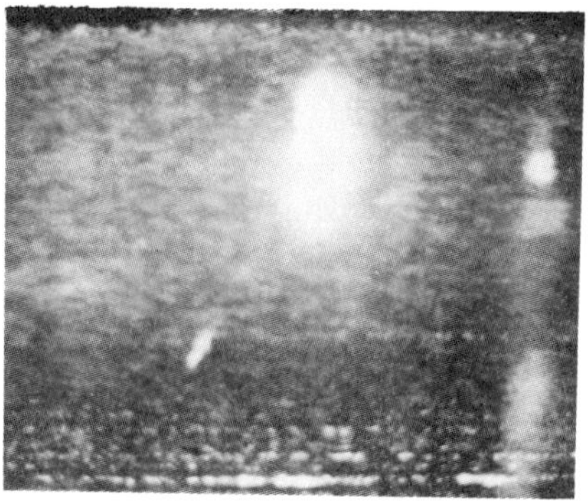

Fig.6-1 (a)

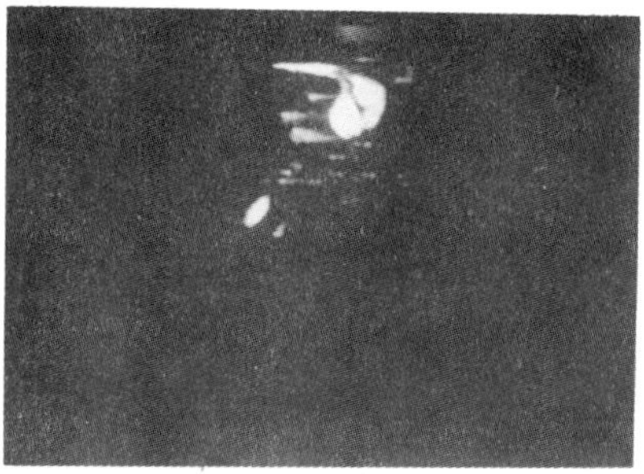

Fig.6-2 (a)

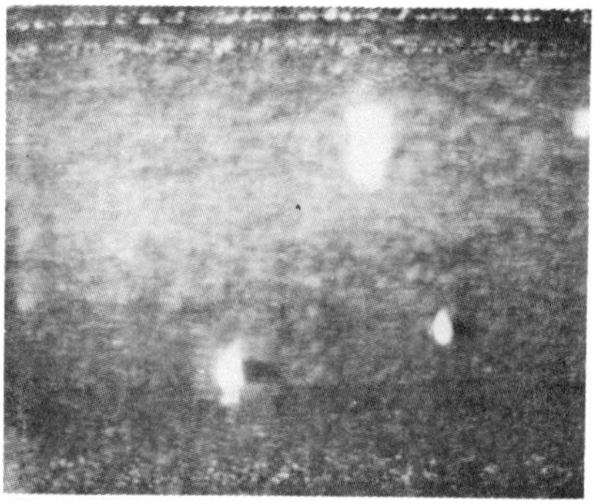

Fig.6-1 (b)

Fig.6-2 (b)

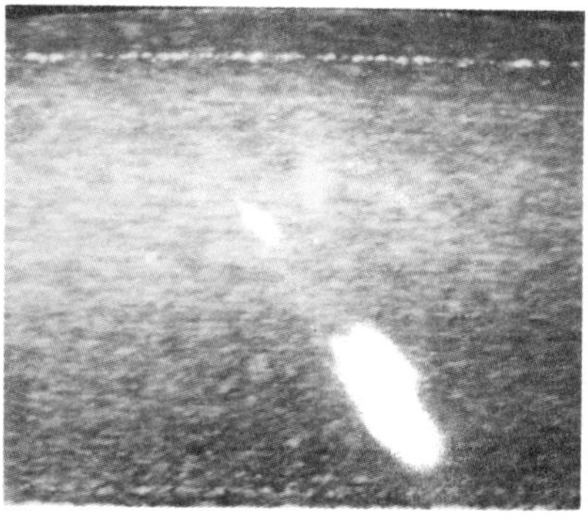

Fig.6-1 (c)

Fig.6-2 (c)

PERFORMANCE OF HIGH-VOLTAGE TEST FACILITY DESIGNED FOR INVESTIGATION OF BALL LIGHTNING.

G.C. Dijkhuis and J. Pijpelink,
Convectron N.V., Rotterdam, The Netherlands.

ABSTRACT.

Along lines specified by the boson model for ball lightning an experimental facility was designed and built for high-voltage breakdown tests in high-velocity flow of a cold gas mixture with controlled chemical composition. A series of blow-down tests without electric discharge demonstrated desired operating conditions for flow parameters and gas composition in the test facility. A series of short-circuit experiments without electric breakdown demonstrated desired operating conditions for charge transfer and electrode voltage from electrical power source and circuit. A series of combined blow-down and breakdown tests showed observed breakdown voltages below calculated threshold voltages for boson collaps in turbulent discharge plasma. Current and voltage profiles observed after breakdown consistently showed phase differences expected from capacitive discharge plasma. Time-resolved pulse counts recorded with a neutron detector stayed below expected levels, and showed unexpected intermittency in signal strength.

1. Introduction.

A recent article (ref. 1) explains long lifetime and high-energy characteristics of some forms of ball lightning as a result of super- conductivity and nuclear fusion in highly turbulent ball lightning plasma. Ref. 1 views ball lightning formation in free air as a phase transition of a free electron gas induced by their mutual exchange interaction . A state diagram specifies the region where free electrons form Cooper pairs and collaps into the collective ground state for particles obeying Bose-Einstein statistics. From a breakdown analysis a bosonisation threshold for atmospheric electric field is obtained that is about four times higher than normal breakdown fields preceding lightning (ref. 2). The structure of a microscopic boson vortex with unit circulation quantum is derived from non-relativistic theory for charged superconductors as a core of self-trapped Cooper pairs surrounded by corotating ions kept in circular orbits by Coulomb forces from the vortex core. Ions are shown to reach fusion velocities when the radius of the vortex core shrinks to about one Bohr radius. As a non-relativistic model, the electron rest energy imposes itself as upper limit of trapped ion energy. The reaction rate of possible fusion processes between atmospheric nuclei in a boson vortex is compared using the Gamow penetration factor for tunneling through a repulsive Coulomb barrier. Ref. 1 concludes that fusion reactions of atmospheric deuterium can supply enough energy to support a ball lightning with sufficient charge trapped inside the boson vortex condensate.

The present article uses the boson model in ref. 1 for design and construction of an experimental facility for investigation of ball lightning. Standard wind tunnel techniques as given in ref. 3 for supersonic Laval nozzles are used to accelerate and cool a gas mixture to flow parameters demanded by the boson model. A standard mixing unit injects a controlled amount of thermonuclear fuel in the mainstream of inert nitrogen gas, beginning at the natural abundance of deuterium in humid air. Standard high-voltage techniques are used for storing charge in a large capacitive power source in quantities comparable to thunderclouds. Standard spark gaps initiate discharge currents at a level comparable to normal lightning .

High- voltage bushings with conventional SF_6-chambers as used in electric power transmission serve to prevent short-circuits through surrounding air. A discharge nozzle with the required combination of strenth and insulation is obtained by fitting a brittle casting with the correct Laval contour inside a pipe wound from strong fibre placed inside a metal housing bolted on opposite sides to both bushings, and in their symmetry plane a pipe leading to the gas supply. On its way to the exhaust in free air, the gas flow passes symmetrical electrodes shaped for maximum breakdown voltage.

For neutron detection a standard gas-filled tube is connected with counting equipment providing digital signal storage during an experiment. A metal cage around the detector serves to shield it from disturbing electromagnetic fields during discharge.

2. Gas flow and mixing.

The gas flow system is shown in Figure 1, showing a tank containing pressurized nitrogen, a gas mixing and conditioning unit, a flow duct and the housing of the discharge nozzle, which is connected to the capacitorbanks that provide the electric power for an experiment.

The pressure of nitrogen gas entering the mixing unit is set with an adjustable throttle valve mounted directly at the outlet of the tank. In practice, this pressure varies between 4 bar and 20 bar, and is effectively constant during one

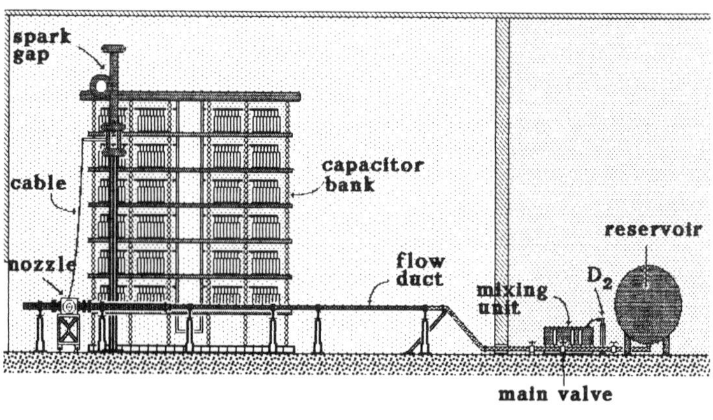

Figure 1: Side-view of Convectron test facility; see text for explanation.

experiment. Deuterium, contained in a gas bottle, is mixed with the nitrogen flow, providing fuel for nuclear fusion reactions. The concentration of the deuterium relative to nitrogen is preset with adjustable flow rate meters. The flow rate meters are set using non-explosive helium gas with nearly the same molecular mass as deuterium. Duration of nitrogen and deuterium flow are set separately with two valves. The main valve controls the nitrogen flow. The valve is opened and closed slow enough to avoid shocks. The gas mixture flows through a pipe with length 23 meters, and reaches the discharge nozzle, contoured as a level nozzle for supersonic flow. At the throat of the nozzle the gas flow is sonic. The nozzle exhausts in free air.

The electrodes, through which the discharge takes place, are positioned in opposite walls of the throat, and are connected by means of two high-voltage bushings and spark gaps to the capacitorbanks.

Pressure is measured first in between the throttle valve and the main valve. The pressure of nitrogen flowing to the mixing unit is set, with the main valve closed. After the gas has passed the main valve the pressure is recorded on a digital memory oscilloscope. Another place for measurement of the pressure is at the throat of the discharge nozzle after removal of one electrode. For this reason the pressure at the throat can't be recorded while doing an experiment.

Conditions of the gas mixture at the throat during a discharge are calculated from a series of test measurements, during which the capacitorbanks are not connected, using the equation of state for nitrogen. These test measurements relate pressure at the throat to pressure at the outlet of the mixing unit.

Figure 2 shows a van der Waals

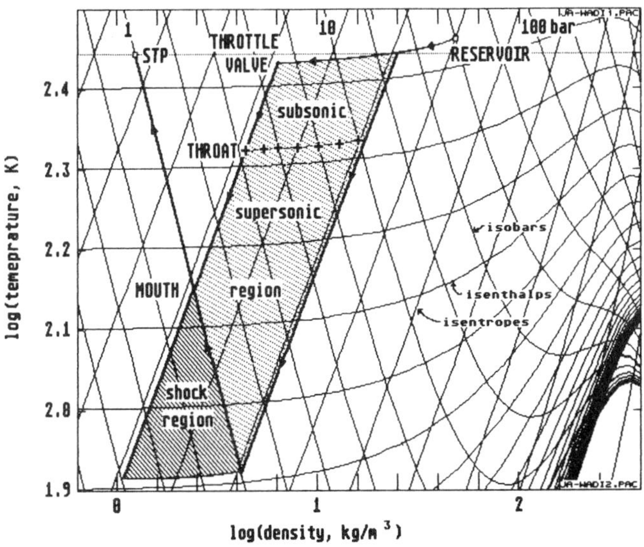

Figure 2: Range of gas flow parameters calculated from van der Waals equation for nitrogen.

diagram for nitrogen, showing the range of pressure, temperature and density in which the experiments are carried out. Some of the experimental settings are indicated in this diagram; the reservoir, having a pressure of 40 bar, is the starting point for all experiments. The gas passes the throttle valve reducing pressure along an isenthalpic path in Figure 2. After having passed the throttle valve, the gas flows through the pipes and the nozzle, along an isentropic track. The pressure at the nozzle throat during a discharge is known from test measurements, and from Figure 2 the temperature and density can be read. As can be seen from Figure 2, most discharges take place in a nitrogen-deuterium mixture, having a temperature somewhat higher then 200 K; the density of the gas mixture varies, dependant upon inlet pressure, preset with the throttle valve.

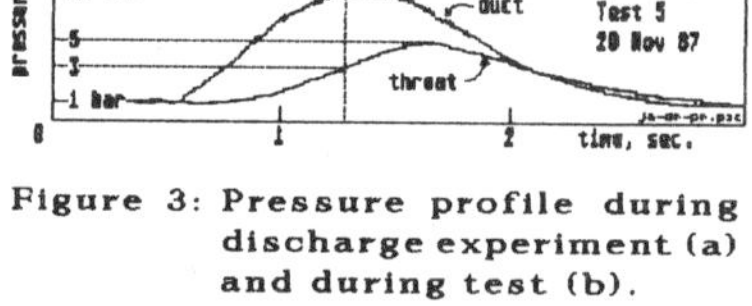

Figure 3: Pressure profile during discharge experiment (a) and during test (b).

During experiments, pressure at the main valve is recorded on a digital storage oscilloscope; the profile in Figure 3 shows a disturbance at the time of the discharge. From test measurements the relation between the pressure at the main valve and at the nozzle throat is known, as shown in Figure 3 for one test. Figure 3 shows throat pressure at the time of discharge estimated from the test measurements and the time of the discharge, known from the disturbance peak of the pressure profile during a breakdown test at KEMA.

In all reported experiments deuterium concentration is set by controlled injection in the main flow during the experiment. The concentration of deuterium, [D], relative to the nitrogen concentration, [N], is calculated afterwards from the preset deuterium flow rate J_d, in normal liters per second, and the nitrogen flow rate J_n, in normal liters per second, from the equation:

$$[D]/[N] = J_d / J_n . \qquad (1)$$

Nitrogen mass flow is calculated in two ways: first from the pressure drop in the gas reservoir observed during one experiment, and the time that the main valve for the nitrogen flow is opened. Second from the assumption that the nozzle is choked. From known throat conditions, the number of moles per second is calculated, and the flow rate in normal liters per second. Both calculations give a mean value for the concentration during the time of the experiment.

Table 1 specifies pressure, temperature, density and flow rate of nitrogen gas at the nozzle throat, for some values of pressure at the throttle valve. The pressure, density and temperature follow from the discharge nozzle being a Mach 2.54 level nozzle; relations for pressure ratio, temperature ratio and density ratio are tabulated in reference 3, table III. In the lower part of table 1 the deuterium abundance is given for several values of the deuterium flow rate; the lower value is not much higher than the natural abundance of deuterium in air, calculated from the vapour pressure

Table 1: Parameters of gas flow for some typical settings.

throttle valve		5 bar	12 bar	18 bar
nitrogen gas at nozzle throat	pressure	2.7 bar	6.4 bar	9.5 bar
	temperature	230 K	230 K	230 K
	density	4 kg/m^3	10 kg/m^3	14 kg/m^3
	flow rate	2.3 m$_n^3$/s	5.6 m$_n^3$/s	8.3 m$_n^3$/s
deuterium abundance at flow rate	0.2 l$_n$/s	86 ppm	36 ppm	24 ppm
	0.4 l$_n$/s	170 ppm	71 ppm	48 ppm
	2.0 l$_n$/s	860 ppm	360 ppm	240 ppm
	4.0 l$_n$/s	1730 ppm	710 ppm	480 ppm

of water, 12.788 mm Hg, at a temperature of 15.0 °C [4]. In nature, one atom out of 6000 hydrogen atoms is a deuterium atom, so 1 out of 3000 water molecules contains a deuterium atom, and the abundance of molecules containing a deuterium atom equals 5.6 parts per million. This means that humid air contains 1.4×10^{20} deuterium atoms per cubic meter. The lower and upper limits of inflammability of hydrogen in air are, according to reference 4, page D-124, equal to 4.00 and 74.20 volume percents respectively.

3. Electrical circuit.

Figure 4 shows a front view of the Convectron test facility, connected to the high-voltage power supply of KEMA. The housing of the discharge nozzle has two high-voltage bushings mounted on it. The bushings contain SF_6-gas at 8 bar, preventing breakdown to the aluminium housing. Copper cables connect these bushings to spark gaps and capacitor banks. The two capacitor-banks are connected in series, with adjustable capacitance set at several microfarads. Each capacitorbank is charged to 240 kilovolts, resulting in a total charge storage around one coulomb. A digital oscilloscope measures voltages reduced with a capacitive voltage divider, not shown in Figure 4. Currents are recorded through Rogowski coils, coupled inductively to the discharge circuit.

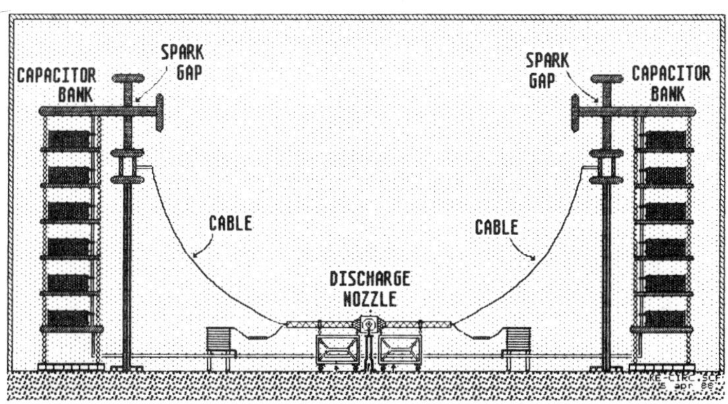

Figure 4: Front view of Convectron test facility.

Current profiles are recorded with a two-channel digital storage ocilloscope for each electrode of the discharge nozzle, and likewise for each voltage profile after breakdown. During tests at KEMA on 21 October, an ADAM Transient Recorder was used for pre-breakdown voltage profiles. This recording equipment was placed outside the test hall shown in Fig. 4. About 50 m of glass fibre cable was used for transmitting current and voltage data from the discharge nozzle to the recording equipment.

Before actual experiments, several test runs without the discharge nozzle are made to check expected voltages and currents. Test runs for the voltage are made with an open circuit, while test runs for the current are made with shorted circuit where the resistance of the connecting cables serves as load. Figure 5 shows the equivalent electrical circuit of the

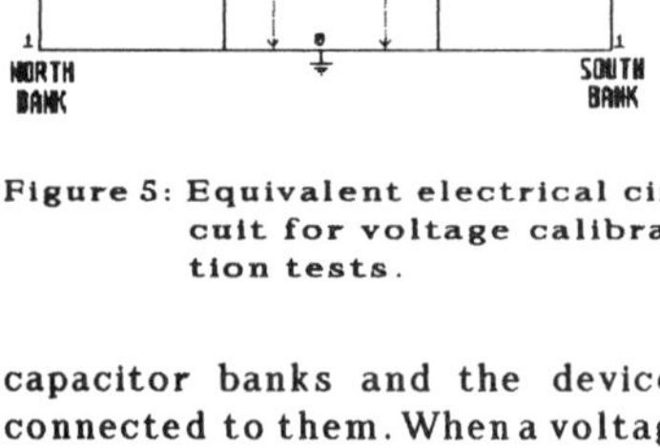

Figure 5: Equivalent electrical circuit for voltage calibration tests.

capacitor banks and the devices connected to them. When a voltage calibration test is made, the gap is opened; for a current calibration test the gap is closed by a copper cable; when a discharge is being done, the discharge nozzle is connected, and the gap is the spacing in between the electrodes inside the discharge nozzle. The circuit consists of two capacitorbanks, each having a capacity of about one microfarad, two spark gaps, with infinite resistance before, and effectively zero resistance after short-circuit, the resistance and inductance of the connecting wires (inductance equals 60 microhenry for each wire; the resistance 0.35 ohm) and some resistors and capacitors, of 12.5 ohm) and several nanofarads, respectively. which are connected to the circuit to symmetrize it. Capacities of the bushings (82 pF each) and the capacitive voltage dividers (200 pF each) can be neglected, relative to the capacity of the capacitorbanks.

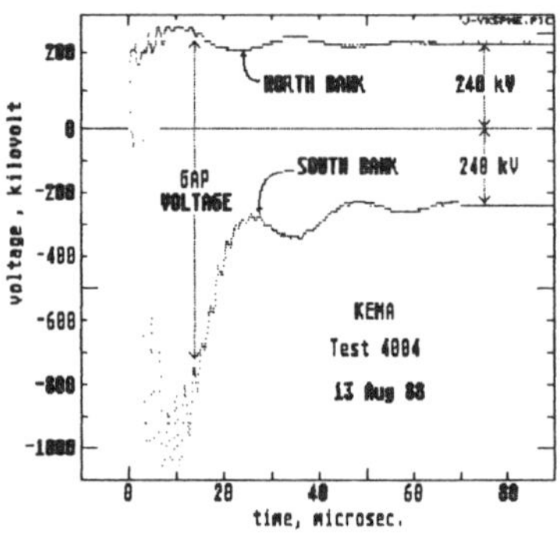

Figure 6: Measured voltage profile during test run.

Figure 6 shows a voltage profile, measured during a voltage test run, showing that the circuit was not symmetrical during this experiment. The measurement shows that voltages up to 1250 kilovolts are within reach of the electrical circuit.

Voltage profiles, calculated with open circuit, are shown in Figure 7. Capacitance C_{h1} is used as parameter in these calculations. These calculations use the method of state variable analysis, as explained in reference 5. The net-

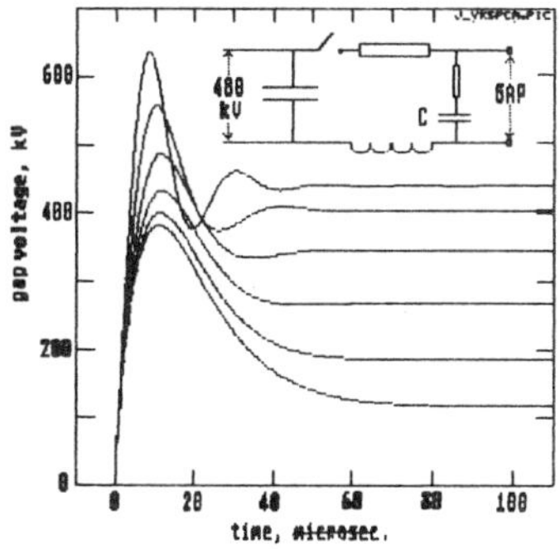

Figure 7: Calculated voltage test profiles.

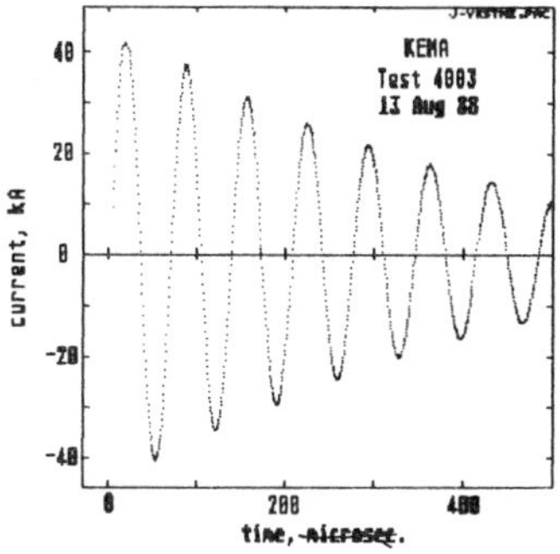

Figure 8: Measured current profile of closed circuit.

work equations are rewritten as simultaneous first-order differential equations, and initial values of capacitor voltages and inductor currents determine their solution. For the circuit shown in Figure 5, the following equations can be written:

$$V_{23} = 0.5*V_0 - V_{12} - V_{50} - I_1*\left(R_{h0} + R_{h1}\right) \qquad (2a)$$

$$I_1 = C_{h0}\frac{dV_{12}}{dt} \qquad (2b)$$

$$V_{23} = L_{h0}\frac{dI_1}{dt} \qquad (2c)$$

$$I_1 = C_{h1}\frac{dV_{50}}{dt} \qquad (2d)$$

$$\left(V_{12}\right)_0 = 0; \left(V_{50}\right)_0 = 0$$

$$\left(I_1\right)_0 = 0; \qquad (2e)$$

$$V_{40} = V_{50} + I_1*R_{h1} \qquad (2f)$$

Integration is done, making use of the Euler method.

Figure 8 shows a measured current profile of a closed circuit. The circuit is nearly the same as the one used for the voltage test runs, the only difference is that the gap, used in the circuit for the voltage test run, is closed with a copper cable, for the current test run. Figure 9 shows some calcula-

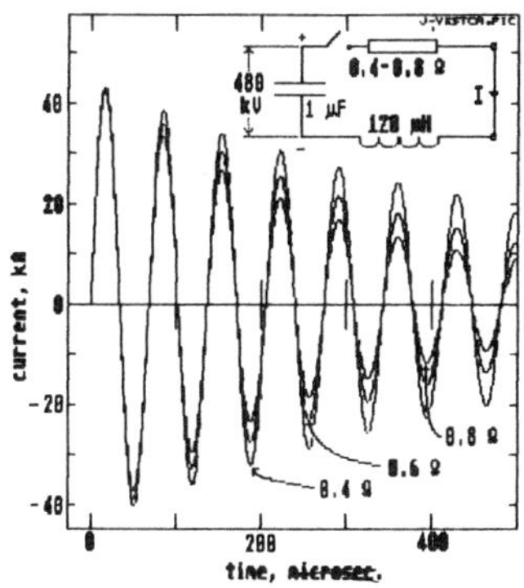

Figure 9: Calculated current profiles for closed circuit.

ted current profiles, for a calibration shot. From these profiles it can be seen that the circuit is a weakly damped series L,C,R-circuit. Short-circuit current profiles are calculated from:

$$I = \frac{V_0}{(L/C - R^2/4)^{1/2}} * \exp\left(\frac{-R}{2L} t\right) *$$

$$* \sin\left\{\left(1/LC - R^2/4L^2\right)^{1/2} * t\right\} \qquad (3)$$

where V_0 is initial voltage of the capacitor banks, L the inductance of the circuit, C the capacity of the capacitor banks and R the resistance of the cables.

For a breakdown experiment, the capacitor banks are charged, the main valve for the gas flow is opened and after that, both spark gaps are short-circuited simultaneously. Voltage at the electrodes inside the discharge nozzle increases, until the breakdown voltage is reached. The magnitude of this breakdown voltage depends upon spacing and shape of the electrodes, and gas density between the electrodes. At breakdown, gas molecules between the electrodes are ionized, and current begins to flow through the discharge region.

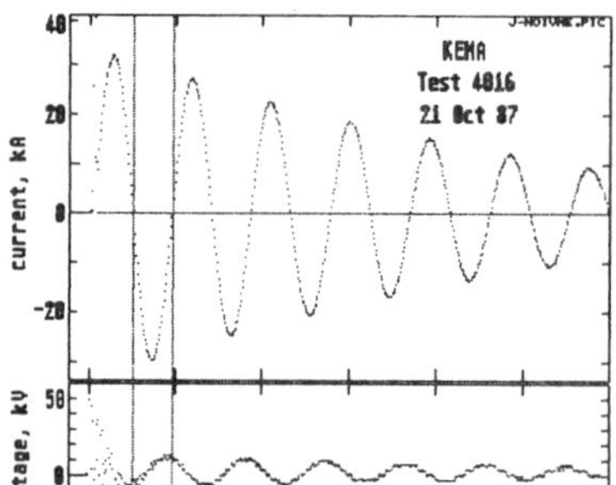

Figure 10: Current and voltage profiles of a discharge.

Figure 10 shows current and voltage profiles of a discharge; the current being measured is flowing through one of the high-voltage bushings. The current reaches a magnitude of 32 kiloamperes; breakdown voltage stayed below 100 kilovolts, the charge stored in the capacitor banks was 0.24 C.

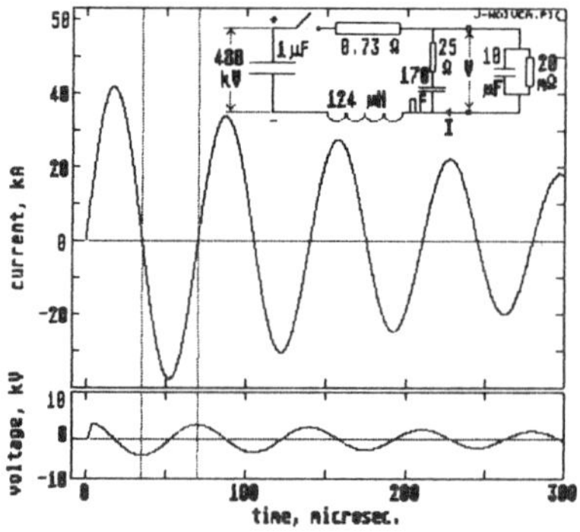

Figure 11: Calculated current and voltage profiles of a discharge.

Figure 11 shows calculated current and voltage profiles of a discharge, together with the circuit in which the calculations are done. The discharge, inside the nozzle, is represented here as a leaky capacitor with parallel capacitance and resistance, values of which are 10 μF and 20 mΩ, respectively. As can be seen in both measured and calculated profiles, current and voltage profiles are out-of-phase; a phase-shift of about 90° exists between voltage and current profiles. The inset of Figure 11 shows an electrical circuit for the experiment together with best values for the circuit components. The period of the oscillation is mainly determined by the capacity of the capacitor banks and the inductance of the wires that connect the spark gaps to the high-voltage bushings. The other components, the spark gaps and the gap inside the discharge nozzle, have little influence on the period of the oscillation. The capacity of the capacitor banks was 0.5 microfarad; the total inductance was 124 microhenry. The period of the oscillation was about

50 microseconds. Measurement of the electrode voltage is found to be more difficult than for the currents. Before breakdown, the voltage rises, in most cases to several hundreds of kilovolts, as recorded after reduction by capacitive voltage dividers. After breakdown however, the voltage across the gap is much lower and very noisy. The signal from the voltage dividers is more difficult to measure. Ideally, the sensitivity of the detection mechanism should be increased an order of magnitude at the moment of the discharge, in a time interval less than one microsecond.

If conditions are right, the discharge is meant to initiate fusion reactions and to form a fire ball. Hence a neutron detector monitors each experiment and a high-speed camera films the outlet of the discharge nozzle.

Table 2: Threshold electron concentration and electric field for entering region of exchange instability according to Equations 3 and 5 of reference 1.

temperature K	concentration 10^{25} el/m^3	electric field kV/cm
200	0.65	1.5
400	2.6	6.0
600	5.8	14
800	10.3	24
1000	16.2	38

Table 2 shows threshold electron concentration and electric field strength, needed for entering the region of exchange instability, calculated from theory given in reference 1, for several values of the absolute temperature. Table 3

Table 3: Threshold electric field and gap voltage for nitrogen at nozzle throat conditions given in Table 1. Highest measured breakdown voltage on 21 October 1987 was 94 kV for KEMA Test 4016.

gas density kg/m^3	electric field kV/cm	gap voltage kV
1.3	17.8	143
4.0	19.2	154
10.0	19.6	157
14.0	19.6	157

shows threshold electric field and gap voltage, with electrode spacing 8 cm, for conditions given in table 1 for nitrogen at the throat of the nozzle. During the experiments at KEMA on 21 October 1987, the highest measured breakdown voltage was 94 kV.

At all experiments carried out at KEMA during October 1987, electric field strength at the moment of breakdown, was lower than the threshold field. These low values for the breakdown voltages resulted from low gas density at the moment of breakdown. However, due to the fact that the electrodes are not two parallel plates, electric field intensity is not constant over the length of the gap; at the electrode surface , the electric field has a larger value, which can be much above the threshold field .

4. Neutron counting.

In the previous section, the fact was mentioned that during an experiment a neutron detector is counting. If conditions during the discharge are right, the discharge initiates a nuclear fusion process.

The process under consideration will be the D-D fusion process, as follows:

$$D + D \rightarrow {}^{3}_{2}He + n + 3.2 \text{ MeV}$$
$$D + D \rightarrow {}^{3}_{1}H + p + 4.2 \text{ MeV} \quad (4)$$

From the first reaction neutrons are emitted with an average energy of about 2.45 MeV.

One way of detecting neutrons is to make use of activation of special materials, such as indium, sulfur or gold, after neutrons are captured by them. The activity of these materials, at a certain time after irradiation by the neutrons, is proportional to the neutron flux. This method of neutron counting however, is much less sensitive than proportional counters; a large neutron dose causes a measurable activation, making use of a gamma radiation detector. A second problem is the short decay time of the activated atoms, varying from one hour to some days. An advantage of this method is that it is totally insensitive for electromagnetic fields, which can disturb a measurement with a detector making use of electronics. During experiments at KEMA in July and August 1987, this method has been applied, without a result. During the experiments in July 1987, no deuterium was injected, due to a disturbance of a control unit for the deuterium valve, while during the experiments in August 1987 the deuterium injection was allright, but the electronics of the neutron detector failed. Because of this, only activation measurements were done during the August experiments. The activation badges were on a distance of about 50 centimeters of the discharge region, so a sphere with an area of 3.14×10^4 cm^2 is irradiated by the neutrons. The minimum amount of neutrons having an energy in the range from 0.3 eV to 1 MeV, that has to be captured by the activation material, is 10^7 cm^{-2}, this resulting in a number of 3.14×10^{11} neutrons that has to be emitted by the source. Any amount of neutrons smaller than this, can't be detected with the activation method. The fact that during these experiments no neutrons are detected by the activation badges, only means that the number of neutrons produced by the discharge was less than the number calculated above. For detection of neutrons during the experiments at KEMA in October 1987, a neutron counter is installed with a moderator tuned to neutrons in this energy range. The neutron-sensitive medium is a BF_3-tube, manufactured by LND. It is a gas-filled tube, with a diameter of two inches, having one anode wire inside. The tube is filled with BF_3-gas at a pressure of 70 cm Hg, which is enriched to 96 % of ^{10}B. The anode wire is insulated from the rest of the tube, and is set at a high voltage. The wall of the tube is grounded. The cross section for neutron absorption for ^{10}B has its largest value for thermal energies. Because of this, neutrons must be moderated before they can be detected effectively with a BF_3 tube. Moderation is done by placing the tube inside a cilinder of parrafin, having a thickness of about 6.5.

When a neutron enters the tube and is captured by a borium atom, the following reaction takes place:

$$^{10}B + n \rightarrow {}^{7}Li + \alpha + 2.4 \text{ MeV} \quad (5)$$

The α-particle ionizes a large amount of gas molecules. Liberated electrons are accelerated by electric field inside the tube, towards the anode wire. In the neighbourhood of the anode wire, the electrical field is strong enough to accelerate free electrons to energies above the ionization threshold of the gas molecules. The electrons are collected on the anode wire, which is connected to the input of a charge-sensitive amplifier.

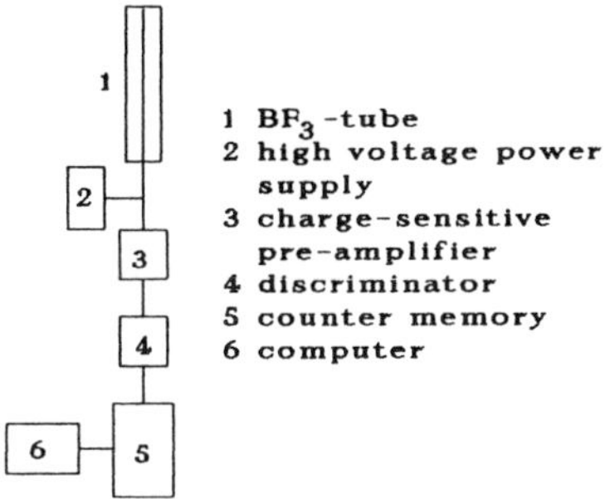

Figure 12: Block diagram of the neutron counter.

A block diagram of the detector is shown in Figure 12. After being amplified, the pulse passes through a discriminator, with a voltage level that stops noise and pulses due to gamma radiation. After this, the signal is converted into a standard TTL-pulse, which is added to an 8-bits counter with a highest count of $2^8 = 256$ pulses. Every half of a millisecond the counter value is read and reset to zero by a Commodore 64 personal computer. One complete duty cycle of the neutron detector comprises 2000 × 0.5 ms = 10 seconds.

To prevent interference from electromagnetic fields, the neutron detector is triggered optically by means of a flash lamp illuminating a photo sensitive cell, some seconds before the discharge takes place. The detector is triggered at the time that the valve for the deuterium injection is opened. After being triggered, the detector is counting for a time of ten seconds. Both detector and computer are shielded for the electromagnetic fields caused by the discharge by placing them inside a Faraday's cage. The only cable outside the Faraday's cage is the one for the power supply for both detector and computer; this cable is connected by means of a transformer that seperates the detector and computer from the electrical mains. The shielding will be more perfect, if no cable at all comes from inside the Faraday's cage outside.

Before using the neutron detector, it is tested and calibrated, using a PuBe-neutron source, emitting neutrons with averaged energy around 5 MeV, not too far from the expected neutron energies. The most suitable moderator thickness was 6.5 cm, and an effective area of about 23 cm^2 for the counting tube was determined. Calibration of the detector as a function of the distance between detector and neutron source, is done in the following way: for several values of the distance between detector and source, the count rate is detected. These measurements are carried out with a moderator of the right thickness around the tube and with no shielding around the neutron source. At the same time, the radiation dose is recorded with a calibrated rem counter. The dose is related to the neutron flux, which stands for the number of neutrons detected per second and per cm^2. From this, and the number of neutrons detected with the detector that is to be calibrated, the effective area is determined. If a number of neutrons is detected, the radiation dose can be calculated, following the opposite direction.

During several test runs at KEMA, the neutron detector was triggered, in order to find out whether the apparatus would be disturbed by the discharge or not. At all test runs, the discharge

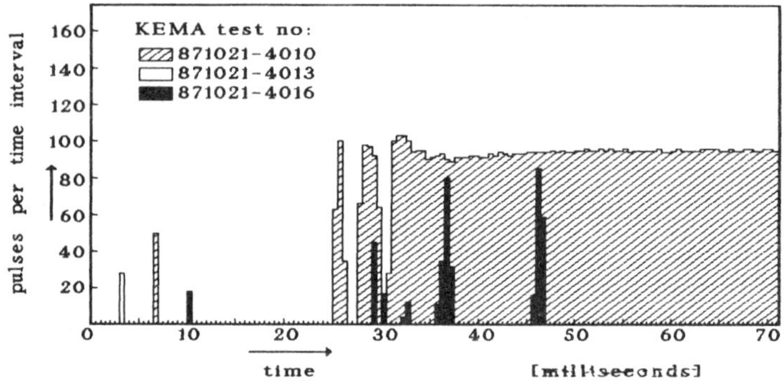

Figure 13: Pulses recorded by the neutron counter for three experiments; one experiment only showing one pulse, a second experiment showing a number of pulses, while during the third experiment the FET was destroyed.

caused a number of counts, inside one time-interval of 0.5 milliseconds. From a number of 7 experiments, carried out at 21 October, 1987, the detector only recorded counts during a time-interval of 0.5 milliseconds for 3 of those experiments. For another 3 experiments, the first peak is recorded again, 18 milliseconds after that followed by a large amount of pulses, which destroyed the field effect transistor at the input of the charge-sensitive amplifier. During the remaining experiment, there are also pulses detected 18 millisecond after the first peak, but here, they didn't destroy the FET. Figure 13 shows the pulses recorded by the neutron detector for all 3 kinds of outcomes. In this figure, time zero equals 2200 milliseconds after triggering of the detector. One of the test runs was intended as a real experiment, but the spark gaps accidentally short-circuited before the gas flow was started. The neutron detector was triggered, probably by sparks of the spark gap, and recorded pulses only during the first period of 0.5 milliseconds; this means that these pulses are caused by the discharge, not only during this measurement, but also during the other experiments. This also means that the pulses that are recorded 18 milliseconds after the discharge, are not caused by the discharge, but something else, that might be initiated by the discharge.

In every measurement made with the neutron detector, whether it is a test run or a real experiment, there is a peak consisting of several pulses during 0.5 milliseconds. During 4 breakdown experiments, out of 7, there are more pulses, in a number of channels, beginning 18 milliseconds after the first peak. In 3 out of 4 cases, the field-effect transistor at the input of the charge-sensitive amplifier is destroyed, something that doesn't happen during 7 test runs.

In future experiments not only neutrons have to be detected, but also the energy spectrum of the neutrons. For this reason a so called neutron spectrum counter is under development. This spectrum counter detects the energy spectrum of the neutrons, as a function of the time; the time interval during which an energy spectrum is recorded, can be set, with a minimum value of one millisecond. This means that every millisecond an energy spectrum is being measured and read by a computer.

During a discharge a high-speed camera filmed the light-effects outside the discharge nozzle. The camera was set for filming a fireball, floating from the nozzle into free air. Although no long-lived fire ball was created, the light-effects during a discharge lasted about ten times longer than the discharge itself.

5. Conclusions:

1. Controll of gas flow and composition as required for the boson model for ball lightning was achieved in the flow channel of the experimental facility described above.

2. Controll of charge storage and gap voltage as required for the boson model was achieved with the electrical circuit unconnected to the discharge nozzle.

3. The discharge nozzle had breakdown voltages at least 40 % below minimal levels specified by the boson model for relevant gas densities .

4. Current and voltage profiles consistently show 90° phase difference as for capacitive discharge plasma behaviour characteristic of the boson model.

5. Pulse counts recorded with a neutron detector during discharge are much lower than expected from the boson model, and show unexpected intermittent behaviour.

References.

1. Dijkhuis, G.C.: "Scaling law for fusion power from boson vortex in ball lightning", Proceedings EMC Symposium, Wroclaw, Poland (June 1988)

2. Berger, K.: "The Earth Flash" in: "Lightning", ed. by R.H. Golde, Volume 1, Academic Press (1977)

3. Liepmann, H.W., and Roshko, A.: "Elements of gasdynamics", John Wiley & Sons (1957)

4. CRC Handbook of Chemistry and Physics, 67th edition, CRC Press (1986-1987)

5. van der Sluis, L.: "A network-independent computer program for the computation of electrical transients", IEEE Transactions on Power Delivary, Vol. PWRD-2, Number 3 (July 1987)

List of Participants

E. Bach (USA)
G. C. Dijikhuis (Zeldenrust College and Convectron N. V., NETHERLANDS)
G. Egely (Hungarian Academy of Sciences, HUNGARY)
F. Fujiki (Tokyo University, JAPAN)
G. P. Gladyshev (USSR Academy of Sciences, USSR)
R. K. Golka (USA)
P. H. Handel (University of Missouri, USA)
H. Hatakeyama (JAPAN)
T. Kaneda (Tokyo Denki University, JAPAN)
A. G. Keul (Universität Salzburg, AUSTRIA)
H. Kikuchi (Nihon University, JAPAN)
N. Kitagawa (Tokyo Household Management University, JAPAN)
P. M. Koloc (Prometheus II. Ltd., USA)
Y. Nagai (Sagami Institute of Technology, JAPAN)
K. Nickel (Albert-Ludwigs-Universität, GERMANY)
H. Ofuruton (Tokyo Metropolitan College of Aeronautical Engineering, JAPAN)
Y. H. Ohtsuki (Waseda University, JAPAN)
K. Shutoh (Science University of Tokyo, JAPAN)
S. Singer (Athenex Research Associates, USA)
B. M. Smirnov (Uspekhi, USSR)
R. Tsuji (Ibaraki University, JAPAN)
Y. H. Tsunoda (JAPAN)
S. Tsurumi (Science University of Tokyo, JAPAN)
K. W. Wevers (Zeldenrust College and Convectron N. V., NETHERLANDS)
H. Yamamoto (Fukui University, JAPAN)
Zou You-Suo (Chinese Academy of Sciences, CHINA)